Communications and Control Engineering

Series Editors

E.D. Sontag · M. Thoma · A. Isidori · J.H. van Schuppen

Published titles include:

Nonlinear Control Systems (Third edition)
Alberto Isidori

Stability and Stabilization of Infinite Dimensional Systems with Applications
Zheng-Hua Luo, Bao-Zhu Guo
and Omer Morgul

Nonsmooth Mechanics (Second edition)
Bernard Brogliato

Nonlinear Control Systems II
Alberto Isidori

L_2-Gain and Passivity Techniques in Nonlinear Control
Arjan van der Schaft

Control of Linear Systems with Regulation and Input Constraints
Ali Saberi, Anton A. Stoorvogel and Peddapullaiah Sannuti

Robust and H_∞ Control
Ben M. Chen

Computer Controlled Systems
Efim N. Rosenwasser and Bernhard P. Lampe

Control of Complex and Uncertain Systems
Stanislav V. Emelyanov and Sergey K. Korovin

Robust Control Design Using H_∞ Methods
Ian R. Petersen, Valery A. Ugrinovski and Andrey V. Savkin

Model Reduction for Control System Design
Goro Obinata and Brian D.O. Anderson

Control Theory for Linear Systems
Harry L. Trentelman, Anton Stoorvogel and Malo Hautus

Functional Adaptive Control
Simon G. Fabri and Visakan Kadirkamanathan

Positive 1D and 2D Systems
Tadeusz Kaczorek

Identification and Control Using Volterra Models
Francis J. Doyle III, Ronald K. Pearson and Bobatunde A. Ogunnaike

Non-linear Control for Underactuated Mechanical Systems
Isabelle Fantoni and Rogelio Lozano

Robust Control (Second edition)
Jürgen Ackermann

Flow Control by Feedback
Ole Morten Aamo and Miroslav Krstić

Learning and Generalization (Second edition)
Mathukumalli Vidyasagar

Constrained Control and Estimation
Graham C. Goodwin, María M. Seron and José A. De Doná

Randomized Algorithms for Analysis and Control of Uncertain Systems
Roberto Tempo, Giuseppe Calafiore and Fabrizio Dabbene

Switched Linear Systems
Zhendong Sun and Shuzhi S. Ge

Subspace Methods for System Identification
Tohru Katayama

Digital Control Systems
Ioan D. Landau and Gianluca Zito

Multivariable Computer-controlled Systems
Efim N. Rosenwasser and Bernhard P. Lampe

Dissipative Systems Analysis and Control (Second edition)
Bernard Brogliato, Rogelio Lozano, Bernhard Maschke and Olav Egeland

Algebraic Methods for Nonlinear Control Systems (Second edition)
Giuseppe Conte, Claude H. Moog and Anna Maria Perdon

Polynomial and Rational Matrices
Tadeusz Kaczorek

Simulation-based Algorithms for Markov Decision Processes
Hyeong Soo Chang, Michael C. Fu, Jiaqiao Hu and Steven I. Marcus

Hyo-Sung Ahn, Kevin L. Moore
and YangQuan Chen

Iterative Learning Control

**Robustness and Monotonic Convergence
for Interval Systems**

 Springer

Hyo-Sung Ahn, PhD
Department of Mechatronics
Gwangju Institute of Science
 and Technology (GIST)
Buk-gu, Gwangju 500-712
Korea

Kevin L. Moore, PhD, PE
Division of Engineering
Colorado School of Mines
Golden, CO 80401
USA

YangQuan Chen, PhD
Center for Self-organizing
 and Intelligent Systems
Utah State University
Logan, UT 84322-4160
USA

British Library Cataloguing in Publication Data
Ahn, Hyo-Sung
  Iterative learning control : robustness and monotonic
  convergence for interval systems. - (Communications and
  control engineering)
  1. Intelligent control systems 2. Iterative methods
  (Mathematics)
  I. Title II. Moore, Kevin L., 1960- III. Chen, Yangquan,
  1966-
  629.8
ISBN-13: 9781846288463

Library of Congress Control Number: 2007928318

Communications and Control Engineering Series ISSN 0178-5354
ISBN 978-1-84628-846-3   e-ISBN 978-1-84628-859-3   Printed on acid-free paper

9 8 7 6 5 4 3 2 1

Springer Science+Business Media
springer.com

This work is dedicated to:

My wife, Min-Hui Kim, and
My parents, Chang-Soo Ahn and Hak-Sun Kim

– Hyo-Sung Ahn

My family, Tamra, Joshua, and Julia, and
Professor Suguru Arimoto

– Kevin L. Moore

The memory of my father Hanlin Chen, and
My family, Huifang Dou, Duyun, David, and Daniel

– YangQuan Chen

Preface

This monograph studies the design of robust, monotonically-convergent iterative learning controllers for discrete-time systems. Iterative learning control (ILC) is well-recognized as an efficient method that offers significant performance improvement for systems that operate in an iterative or repetitive fashion (e.g., robot arms in manufacturing or batch processes in an industrial setting). Though the fundamentals of ILC design have been well-addressed in the literature, two key problems have been the subject of continuing research activity. First, many ILC design strategies assume nominal knowledge of the system to be controlled. Only recently has a comprehensive approach to robust ILC analysis and design been established to handle the situation where the plant model is uncertain. Second, it is well-known that many ILC algorithms do not produce monotonic convergence, though in applications monotonic convergence can be essential. This monograph addresses these two key problems by providing a unified analysis and design framework for robust, monotonically-convergent ILC.

The particular approach used throughout is to consider ILC design in the iteration domain, rather than in the time domain. Using a lifting technique, the two-dimensional ILC system, which has dynamics in both the time and iteration domains, is transformed into a one-dimensional system, with dynamics only in the iteration domain. The so-called super-vector framework resulting from this transformation is used to analyze both robustness and monotonic convergence for typical uncertainty models, including parametric interval uncertainties, frequency-like uncertainty in the iteration domain, and iteration-domain stochastic uncertainty. For each of these uncertainty models, corresponding ILC design concepts are developed, including optimization-based strategies when the system Markov matrix is subject to interval uncertainties, an algebraic approach to iteration-domain H_∞ control, and iteration-domain Kalman filtering.

Readers of the monograph will learn how parametric interval concepts enable the design of less conservative ILC controllers that ensure monotonic convergence while considering all possible interval model uncertainties. Additionally, by considering H_∞ techniques and Kalman filtering in the iteration domain using the super-vector framework, the notion of ILC baseline error is established analytically, leading the reader to understand the fundamental limitations of ILC.

The monograph is organized into three different parts and an appendix section. In Part I, we provide research motivations and an overview of the literature. This part of the monograph gives an introduction to ILC and introduces the basic robustness and monotonic convergence issues that can arise. A brief summary of the ILC literature is given and the super-vector approach is presented. In Part II, the concept of a parametric interval is used to reduce the conservatism arising in testing the robust stability of traditional robust ILC methods. In this part, Markov vertex matrices are used for analyzing the monotonic convergence of uncertain ILC systems and, based on this analysis, two different synthesis methods are developed to design ILC controllers that provide robust stability for such systems. In addition to analysis and synthesis of ILC laws for interval systems, it is shown how to develop suitable Markov interval models from state-space interval models. In Part III, the concepts of H_∞ ILC, Kalman filter-augmented ILC, iteration-varying robust ILC, and intermittent ILC are developed. In H_∞ ILC, a unified robust framework for handling model uncertainty, iteration-varying disturbances, and stochastic noise is developed. In Kalman filter-augmented ILC, the baseline ILC error is analytically calculated for use in the off-line design of an ILC update algorithm. Next we consider ILC algorithm design for the case of plants with iteration-varying model uncertainty, under the requirement of monotonic convergence. We conclude this part by considering a Kalman filter-based ILC design for the case where the system experiences data dropout. There are four appendices, comprising a taxonomy of ILC literature published since 1998 and three separate studies, each introducing and solving a fundamental interval computation problem: finding the maximum singular value of an interval matrix, determining the robust stability of an interval polynomial matrix, and obtaining the power of an interval matrix. These three solutions are used as basic tools for designing robust ILC controllers.

No work can be fully credited to its authors, as we all depend on the support and contributions of those around us. Thus, the first author would like to express his appreciation to his parents, Mr. Chang-Soo Ahn and Mrs. Hak-Sun Kim, for their support while he was completing this monograph. It is impossible to express the greatest conceivable gratitude that is owed to his wife, Min-Hui Kim, for her sacrifice, patience, and understanding throughout his studies and his research. Without her support and love, he could not have completed this monograph. Likewise, the second author would like to thank his family, Tamra, Joshua, and Julia, for allowing him to pursue an academic career and a research lifestyle, and Professor Suguru Arimoto, for his inspiring contributions to the field of iterative learning control. Finally, the third author would like to thank his family, Huifang Dou, Duyun, David, Daniel for their patience and understanding. We would also like to thank the able team at Springer-Verlag, especially Frank Holzwarth, a true TeX-pert who helped us solve a number of typesetting problems, Sorina Moosdorf, who helped us with formatting issues, and our editor Oliver Jackson, whose interest in our project greatly contributed to its successful completion.

Gwangju, Korea   *Hyo-Sung Ahn*
Golden, Colorado, USA   *Kevin L. Moore*
Logan, Utah, USA   *YangQuan Chen*
                                                                 May 2007

Acknowledgments

The purpose of this book is to give a single, unified presentation of our research, based on a series of papers and articles that we have written and on the dissertation of the first author. To achieve this goal, it has been necessary at times to reuse some material that we previously reported in various papers and publications. Although in most instances such material has been modified and rewritten for the monograph, copyright permission from several publishers is acknowledged as follows:

Acknowledgment is given to the Institute of Electrical and Electronic Engineers (IEEE) for permission to reproduce material from the following papers.

©2005 IEEE. Reprinted, with permission, from Hyo-Sung Ahn, Kevin L. Moore, YangQuan Chen, "Schur stability radius bounds for robust iterative learning controller design," *Proc. of the 2005 American Control Conference*, Portland, OR, Jun. 8–10, 2005, pp. 178–183 (material found in Chapter 5).

©2005 IEEE. Reprinted, with permission, from Hyo-Sung Ahn, Kevin L. Moore, YangQuan Chen, "Monotonic convergent iterative learning controller design based on interval model conversion," *Proc. of the 2005 IEEE Int. Symposium on Intelligent Control*, Limassol, Cyprus, Jun. 27–29, 2005, pp. 1201–1206 (material found in Chapter 6).

©2005 IEEE. Reprinted, with permission, from Kevin L. Moore, YangQuan Chen, Hyo-Sung Ahn, "Algebraic H_∞ design of higher-order iterative learning controllers," *Proc. of the 2005 IEEE Int. Symposium on Intelligent Control*, Limassol, Cyprus, Jun. 27–29, 2005, pp. 1207–1212 (material found in Chapter 7).

©2005 IEEE. Reprinted, with permission, from Hyo-Sung Ahn, Kevin L. Moore, YangQuan Chen, "Monotonic convergent iterative learning controller design with iteration varying model uncertainty," *Proc. of the 2005 IEEE Int. Conference on Mechatronics and Automation*, Niagara Falls, Canada, July 2005, pp. 572–577 (material found in Chapter 8).

©2006 IEEE. Reprinted, with permission, from Hyo-Sung Ahn, Kevin L. Moore, YangQuan Chen, "Monotonic convergent iterative learning controller design based on interval model conversion," *IEEE Trans. on Automatic Control*, Vol. 51, No. 2, pp. 366–371, 2006 (material found in Chapter 6).

©2006 IEEE. Reprinted, with permission, from Hyo-Sung Ahn, Kevin L. Moore, YangQuan Chen, "Kalman filter augmented iterative learning control on the iteration domain," *Proc. of the 2006 American Control Conference*, Minneapolis, MN, June 14–16, 2006, pp. 250–255 (material found in Chapter 8).

©2006 IEEE. Reprinted, with permission, from Hyo-Sung Ahn, YangQuan Chen, Kevin L. Moore "Maximum singular value and power of an interval matrix," *Proc. of the 2006 IEEE Int. Conference on Mechatronics and Automation*, Luoyang, China, June 25–28, 2006, pp. 678–683 (material found in Appendix B and Appendix D).

©2006 IEEE. Reprinted, with permission, from Hyo-Sung Ahn, Kevin L. Moore, YangQuan Chen, "A robust Schur stability condition for interval polynomial matrix systems," *Proc. of the 2006 IEEE Int. Conference on Mechatronics and Automation*, Luoyang, China, June 25–28, 2006, pp. 672–677 (material found in Appendix C).

©2006 IEEE. Reprinted, with permission, from Hyo-Sung Ahn, YangQuan Chen, Kevin L. Moore, "Intermittent iterative learning control," *Proc. of the 2006 IEEE Int. Symposium on Intelligent Control*, Munich, Germany, Oct. 4–6, 2006, pp. 832–837 (material found in Chapter 8).

©2007 IEEE. Reprinted, with permission, from Hyo-Sung Ahn, YangQuan Chen, Kevin L. Moore, "Iterative learning control: brief survey and categorization," *IEEE Trans. on Systems, Man, and Cybernetics, Part-C*, Vol. 37, 2007 (material found in Chapter 2 and Appendix A).

©2007 IEEE. Reprinted, with permission, from Hyo-Sung Ahn, YangQuan Chen, "Exact maximum singular value calculation of an interval matrix," *IEEE Trans. on Automatic Control*, Vol. 52, No. 3, pp. 510–514, March 2007 (material found in Appendix B).

Acknowledgment is given to the International Federation of Automatic Control (IFAC) for permission to reproduce material from the following papers:

Hyo-Sung Ahn, Kevin L. Moore, YangQuan Chen, "Stability analysis of discrete-time iterative learning control systems with interval uncertainty," *Automatica*, Vol. 43, No. 5, pp. 892–902, May 2007 (material found in Chapter 4).

Hyo-Sung Ahn, Kevin L. Moore, YangQuan Chen, "Stability analysis of iterative learning control system with interval uncertainty," *Proc. of the 16th IFAC World Congress*, Prague, Czech Republic, July 4–8, 2005 (material found in Chapter 4).

Acknowledgment is given to John Wiley & Sons for permission to reproduce material from the following paper:

"Iteration domain H_∞−optimal iterative learning controller design," Kevin L. Moore, Hyo-Sung Ahn, YangQuan Chen, accepted to appear (in 2008) in *International Journal of Nonlinear and Robust Control* (material found in Chapter 7), ©2007 John Wiley & Sons Ltd.

Contents

Part I

Iterative Learning Control Overview

1

Introduction

1.1 General Overview of the Monograph

This monograph focuses on robust monotonically-convergent iterative learning control (MC-ILC) systems and stochastic iterative learning control systems. For robust MC-ILC design, parametric interval uncertainty models are considered. For stochastic ILC design, norm-bounded uncertainty, external disturbances, stochastic measurement noise, and intermittent measurements are considered, all in the iteration domain.

The monograph is organized into three parts and an appendix section. In the first part of the monograph, the ILC problem is described and motivated in Chapter 1. Then Chapter 2 gives an overview of the ILC literature, covering specifically the literature published between 1998 and 2004. In Chapter 3 the super-vector ILC (SVILC) framework is introduced for use in Chapters 4 to 8. The second part of the monograph considers interval ILC analysis (Chapter 4) and interval ILC synthesis (Chapter 5 and Chapter 6). The focus in the second part of the monograph is on plants with parametric interval uncertainty models. In addition to analysis and synthesis for such systems, it is shown how to develop suitable Markov interval models from state-space interval models. The third part of this monograph discusses H_∞ SVILC design (Chapter 7) and stochastic ILC (Chapter 8). The focus in the third part of the monograph is on asymptotic stability and monotonic convergence conditions of ILC systems under assumptions of iteration-domain model uncertainty and stochastic disturbances and noise. In the Appendix section, a taxonomy of the ILC literature is presented and three fundamental interval computational problems are introduced and solved. Although these interval problems were initially motivated for solving interval ILC problems, due to their own completeness and their potential impact on robust control research in general, these results are carefully described in the appendices.

1.2 Iterative Learning Control

1.2.1 What Is Iterative Learning Control?

Control systems have played an important role in the development and advancement of modern civilization and technology. Control problems arise in practically all engineering areas and have been studied both by engineers and by mathematicians. Industrially, control systems are found in numerous applications: quality control in manufacturing systems, automation, network systems, machine tool control, space engineering, military systems, computer science, transportation systems, robotics, social systems, economic systems, biological/medical engineering, and many others. Mathematically, control engineering includes modeling, analysis, and design aspects. The key feature of control engineering is the use of feedback for performance improvement of a controlled dynamic system. The branches of modern control theories are broad and include classical control, robust control, adaptive control, optimal control, nonlinear control, neural networks, fuzzy logic, and intelligent control, with each branch being distinguished from the others based on the assumptions made about the properties of the systems to be controlled and the performance objectives of the specific methodology under consideration.

Iterative Learning Control (ILC) is one of the more recent control theories. ILC, which can be categorized as an intelligent control methodology,[1] is an approach for improving the transient performance of systems that operate repetitively over a fixed time interval. Although control theory provides numerous tools for improving the performance of a dynamic system, it is not always possible to achieve a desired level of performance. This may be due to the presence of unmodeled dynamics, parametric uncertainties, or disturbances and measurement noise exhibited during actual system operation. The inability to achieve a desired performance may also be due to the lack of suitable design techniques [298]. In particular, when the system is nonlinear, it is not easy to achieve perfect tracking using traditional control theory. However, for a specific class of systems – those that operate repetitively – ILC is a design tool that can help overcome the shortcomings of traditional controllers, making it possible to achieve perfect tracking or performance when there is model uncertainty or when we have a "blind" system.[2]

---

[1] From "Defining Intelligent Control – Report of the Task Force on Intelligent Control," IEEE Control Systems Society, Panos Antsaklis, Chair, Dec., 1993: "Intelligent control uses conventional control methods to solve lower level control problems ... conventional control is included in the area of intelligent control. Intelligent control attempts to build upon and enhance the conventional control methodologies to solve new, challenging control problems."

[2] "Blind" means we have little or no information about the system structure and its nonlinearities. We can only measure input/output signals such as voltage, velocity, position, etc.

Various definitions of ILC have been given in the literature. Some of them are quoted here.

- "The learning control concept stands for the repeatability of operating a given objective system and the possibility of improving the control input on the basis of previous actual operation data." Arimoto, Kawamura, and Miyazaki [29].
- ILC is a "... recursive online control method that relies on less calculation and requires less a priori knowledge about the system dynamics. The idea is to apply a simple algorithm repetitively to an unknown plant, until perfect tracking is achieved." Bien and Huh [43].
- "Iterative learning control is an approach to improving the transient response performance of a system that operates repetitively over a fixed time interval." Moore [298].
- "Iterative learning control considers systems that repetitively perform the same task with a view to sequentially improving accuracy." Amann, Owens, and Rogers [10].
- "ILC is to utilize the system repetitions as experience to improve the system control performance even under incomplete knowledge of the system to be controlled." Chen and Wen [66].
- ILC is a "... controller that learns to produce zero tracking error during repetitions of a command, or learns to eliminate the effects of a repeating disturbance on a control system output." Phan, Longman, and Moore [363].
- "The main idea behind ILC is to iteratively find an input sequence such that the output of the system is as close as possible to a desired output. Although ILC is directly associated with control, it is important to note that the end result is that the system has been inverted." Markusson [287].

All definitions about ILC will have their own emphases. However, a common feature of the definitions above is the idea of "repetition." Learning through a *predetermined* hardware repetition is the key idea of ILC. Hardware repetition can be thought of as a physical layer on the uniformly distributed time axis for providing experience to the mental layer of ILC. "Predetermined" means that the ILC system requires some postulates that define the learning environment of a control algorithm. A person learns about his/her living environment by experience, where the physical layer is the daily activity and the mental layer is the memory of strongly perceived events that are closely related with his/her interest. These strongly perceived events of the past provide knowledge to the human being that can be used to inform the person's current activity. In ILC, the current activity is a control force and the past experience is stored data. A difference between human learning and machine learning is in "predetermined." For a human being, knowledge by learning could be based on similarity and impression, whereas in a machine the initial setup, fixed time point, uniform sampling, repetitive desired trajectory, etc. are predetermined, which then determines the future of the hardware machine. Thus, ILC is concerned with the problem of refining the input to a system that operates repetitively, so that the future behavior of the system is predetermined to improve its current operation over its past operation through the use of past experience.

Consider a system in an initial state to which a fixed-length input signal is applied. After the complete input has been applied, the system is returned to its initial state and the output trajectory that resulted from the applied input is compared to a desired trajectory. The error is used to construct a new input signal (of the same length) to be applied the next time the system operates. This process is then repeated. The goal of the ILC algorithm is to properly refine the input sequence from one trial to the next trial so that as more and more trials are executed the output will approach the desired trajectory.

The basic idea of ILC is illustrated in Figure 1.1. Standard assumptions are that the plant has stable dynamics or satisfies some kind of Lipschitz condition, that the system returns to the same initial conditions at the start of each trial, that the trial lasts for a fixed time T_f, and that each trial has the same length. A typical ILC algorithm for the architecture depicted in Figure 1.1 has the form

$$u_{k+1}(t) = u_k(t) + \gamma \frac{d}{dt} e_k(t), \tag{1.1}$$

where $u_k(t)$ is the system input and $e_k(t) = y_d(t) - y_k(t)$ is the error on trial k, with $y_k(t)$ the system output and $y_d(t)$ the desired response. For a large class of systems this algorithm can be shown to converge in the sense that as $k \to \infty$ we have $y_k(t) \to y_d(t)$ for all $t \in [0, T_f]$. Notice that this algorithm is noncausal. To see this more clearly, note that a discrete-time version of (1.1) can be given as

$$u_{k+1}(t) = u_k(t) + \gamma e_k(t+1), \tag{1.2}$$

where now t is an integer. Clearly (1.2) is noncausal with respect to time. This is a key feature of ILC. Though the algorithm actually acts only on past data, the fact that the initial conditions are reset at the beginning of each trial allows us to do "noncausal" processing on the errors from the previous trial.

Based on the ILC system definition as depicted in Figure 1.1 and following the definitions quoted above, we will propose the following definition:

> ILC is an approach to improve the transient response performance of an unknown or uncertain hardware system that operates repetitively over a fixed time interval by eliminating the effects of a repeating disturbance and by using the previous actual operation data.

Finally, having defined ILC, it is important to point out the focus of ILC research, as clearly defined in the following quote:

- "We learned that ILC is about enhancing a system's performance by means of repetition, but we did not learn how it is done. This brings us to the

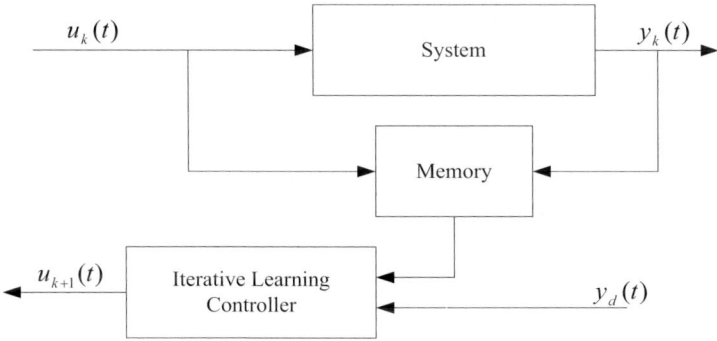

Fig. 1.1. Basic idea of ILC

core activity in ILC research, which is the construction and subsequent analysis of algorithms." Verwoerd [467].

Thus, the key question of ILC is how to eliminate the periodic disturbance and how to use the past information for the current trial. As we will see below, if the system uncertainty and external disturbances are predetermined on the uniformly distributed repetitive time axis, all of these effects, including the actual plant, could be considered as a predetermined model, so that finding an inverse of the deterministic system will be the main objective of ILC.

1.2.2 Classical ILC Update Law

As shown from the taxonomy given in Appendix A, the scope of ILC research is so wide that it is nearly impossible to describe all the branches of ILC. Thus, in this subsection we review only the basic ideas of classical ILC algorithms. Let us consider the following simple SISO linear repetitive system, in continuous time:

$$\dot{x}_k(t) = Ax_k(t) + Bu_k(t) \tag{1.3}$$
$$y_k(t) = Cx_k(t). \tag{1.4}$$

The control task is to servo the output $y_k(t)$ to track the desired output $y_d(t)$ for all time t as the iteration k increases. In classical ILC, the following basic postulates are required, although in recent ILC research, algorithms are sought so that these postulations could be somewhat broken or relaxed (adopted from page 2 of [66]):

- Every trial (pass, cycle, batch, iteration, repetition) ends in a fixed time of duration.
- Repetition of the initial setting is satisfied. That is, the initial state $x_k(0)$ of the objective system can be set to the same point at the beginning of each iteration.

- Invariance of the system dynamics is ensured throughout the repetition.
- The output $y_k(t)$ is measured in a deterministic way.

To give a flavor of ILC results, consider the learning control algorithm proposed in 1984 by Arimoto (and hence, called an "Arimoto-type" ILC law) [27, 28]:

$$u_{k+1}(t) = u_k(t) + \Gamma \dot{e}_k(t). \tag{1.5}$$

For this algorithm and the plant in (1.3–1.4), convergence is assured if

$$\|I - CB\Gamma\|_i < 1. \tag{1.6}$$

Arimoto also considered more general "PID-type" ILC algorithms of the form:

$$u_{k+1} = u_k + \Phi e_k + \Gamma \dot{e}_k + \Psi \int e_k dt. \tag{1.7}$$

In other types of algorithms, researchers have used gradient methods to optimize the gain G_k in:

$$u_{k+1}(t) = u_k(t) + G_k e_k(t+1). \tag{1.8}$$

For ILC design purposes it is sometimes useful to specify the learning control algorithm in the frequency domain, for example:

$$U_{k+1}(s) = L(s)[U_k(s) + aE_k(s)]. \tag{1.9}$$

Note that in this last case the ILC gain is actually implied to be a linear time-invariant filter.

Many schemes in the literature can be classified with one of the algorithms given above. As such, it is possible to generalize the analysis by introducing an operator-theoretic notation. Let $T(\cdot)$ denote a general operator mapping an input space to an output space. Then the following theorem summarizes a number of technical convergence results for first-order[3] ILC systems:

Theorem 1.1. [298] *For the plant $y_k = T_s u_k$, the linear time-invariant learning control algorithm*

$$u_{k+1} = T_u u_k + T_e(y_d - y_k) \tag{1.10}$$

converges to a fixed point $u^*(t)$ given by

$$u^*(t) = (I - T_u + T_e T_s)^{-1} T_e y_d(t) \tag{1.11}$$

with a final error

---

[3] First-order ILC means that only data from the most recent (previous) trial is used in the ILC update law.

$$e^*(t) = \lim_{k \to \infty} (y_k - y_d) = (I - T_s(I - T_u + T_e T_s)^{-1} T_e) y_d(t) \qquad (1.12)$$

defined on the interval (t_0, t_f), *if*

$$\|T_u - T_e T_s\|_i < 1. \qquad (1.13)$$

∎

Note that if $T_u = I$ then $\|e^*(t)\| = 0$ for all $t \in [t_o, t_f]$; otherwise the final converged error will be nonzero.

To understand the nature of ILC, consider the following:

Question: Given T_s, for the general linear ILC algorithm:

$$u_{k+1}(t) = T_u u_k(t) + T_e(y_d(t) - y_k(t)),$$

how do we pick T_u and T_e to make the final error $e^*(t)$ as "small" as possible?

Answer: Let the solution of the following problem be T_n^*:

$$\min_{T_n} \|(I - T_s T_n) y_d\|.$$

It turns out that we can specify T_u and T_e in terms of T_n^* and the resulting learning controller converges to an optimal system input given by:

$$u^*(t) = T_n^* y_d(t).$$

This means that the essential effect of a properly designed learning controller is to produce the output of the best possible inverse of the system in the direction of y_d [298]. This is the key characteristic of ILC.

Returning to the classic Arimoto-type ILC law (1.5), note that the basic formula for selecting the learning gain given in (1.6) does not require information about the system matrix A, which implies that ILC is an effective control scheme for improving the performance of uncertain (linear or nonlinear) dynamic systems. This is the main feature of ILC, as distinguished from classical control theories.

The ILC update rule of (1.5) is properly called a "D-type" ILC rule, as it operates on the derivative of the error. Likewise, we can consider PID-type ILC as given in (1.7), I-type ILC , or P-type ILC. For instance a P-type update rule (meaning no derivative and integral effects) can be written as

$$u_{k+1}(t) = u_k(t) + \gamma_k(t)(y_d(t) - y_k(t)), \qquad (1.14)$$

where k is the iteration trial, $\gamma_k(t)$ is the proportional learning gain, $y_d(t)$ is the desired output, and $y_k(t)$ is the measured output. Note that this particular algorithm also has another feature: the gain is time-varying. This introduces another way to categorize ILC update laws: as time-invariant or time-varying.

As we noted above, we can also consider learning gains that are filters (e.g., the update law in (1.9)), which leads us to also consider the distinction between time-varying and time-invariant learning gain filters.

As we mentioned in a footnote above, ILC that only looks back at the most recent previous iteration is called first-order ILC. It is also possible to consider what is called higher-order ILC (HOILC), whereby data from more than one previous iteration is used. Consider, for instance, the ILC law

$$u_{k+1}(t) = u_k(t) + \sum_{i=k}^{i=k-l} \gamma_i(t)(y_d(t) - y_i(t)) \tag{1.15}$$

or the update algorithm

$$u_{k+1}(t) = \sum_{i=k}^{i=k-l} \lambda_i(t)u_i(t) + \sum_{i=k}^{i=k-l} \gamma_i(t)(y_d(t) - y_i(t)). \tag{1.16}$$

Similar to (1.15) and (1.16), which are both P-type rules, I-type, PD-type, and PID-type higher-order algorithms can be developed using the whole set of past control input signals and output error signals. [66] formulates perhaps the most general form of these type of algorithms, which is given as

$$u_{k+1} = \sum_{k=1}^{N}(I - \Lambda)P_k u_k + \Lambda u_o$$

$$+ \sum_{k=1}^{N}\left(\Phi_k e_{i-k+1} + \Gamma_k \dot{e}_{i-k+1} + \Psi_k \int e_{i-k+1}\mathrm{d}t\right). \tag{1.17}$$

It can be shown [66] that if $\sum_{k=1}^{N} P_k = I$, then by properly choosing the learning gain matrices we can ensure that e_k converges to zero asymptotically.

So far, we have considered continuous time ILC algorithms. However, practically it is desirable to use the discrete-time system state-space model given below in (1.18–1.20), because microprocessor-based discrete, sampled-data systems are widely used in actual applications. Furthermore, since the nature of repetitive operations is finite-horizon, each iteration domain consists of finite number of discrete-time points, which can be represented by vectors (see (1.23–1.26) below). Thus, in the remainder of this monograph we will restrict our attention almost exclusively to discrete-time systems.

1.2.3 The Periodicity and Repetitiveness in ILC

In the descriptions of ILC given above, it has been implied that repetition in the system operation is with respect to time. However, more generally, the periodicity and the repetitiveness treated in ILC could be time-, state-, iteration-, or trajectory-dependent. Let us consider Figure 1.2(a) where the mobile wheel is misaligned with the robot body. In this figure, the robot body Y axis (Y_b) is not pointing in the same direction as the wheel Y axis (Y_w). This misalignment (δ) results in an eccentricity problem in the control system, which brings about an angle-dependent periodic disturbance. Thus, the eccentricity (f_e) is a function of angle θ, i.e., $f_e = f(\theta)$. Figure 1.2(b) shows a satellite that rotates around the earth periodically. For a specified mission, the satellite is controlled to point in a particular direction. However, the satellite experiences external disturbances in the control system from the earth, sun, magnetic field, solar radiation, etc. These disturbances would be time-periodic, because the satellite orbit period is generally fixed. Figure 1.2(c) shows a robot manipulator which is time-periodic and whose initial condition at the start of each new period is same. The robot manipulator works time-periodically on a fixed trajectory (in this figure, by an angle θ), but experiences iteration-varying disturbances. Thus, as shown in these figures, a periodic system could be defined as being in one of three different classes:

- Class A: state-periodic, but not time-periodic (Figure 1.2(a)).
- Class B: time-periodic, but not state-periodic (Figure 1.2(b)).
- Class C: state (starting state)-periodic and time-periodic (Figure 1.2(c)).

In more detail, in Figure 1.2(a), the eccentricity is the function of θ with the following relationship: $f_e(\theta) = f_e(\theta \pm 2n\pi)$ where n is an integer; in Figure 1.2(b), the external disturbance f_d is the function of t with the following relationship: $f_d(t) = f_d(t \pm nT)$ where T is the orbit period; in Figure 1.2(c), the external disturbance could be time-periodic but with the same initial states. Generally, iterative learning control treats the time-periodic system like Class C, whereas the periodic systems like Class A and Class B are studied under repetitive control (RC) and/or periodic control.[4] In this monograph, we will focus on Class C, although our framework can be easily modified to cover Class A and Class B.

1.2.4 Advantages of Using ILC

In the previous subsections, we discussed the characteristics of ILC, introduced basic ILC algorithms, and looked at different ways periodicity can occur. In this subsection, we present some advantages of ILC over typical control algorithms. These include:

- Precise trajectory tracking: If the four postulations given in Section 1.2.2 are satisfied, then a desired trajectory on a finite horizon in the time domain can be perfectly achieved. Thus, ILC algorithms can be effectively used for precise control in fields like semiconductor manufacturing

---

[4] However, Class A and Class B also have been widely studied in ILC. More or less, these days there is no distinction between the systems treated in ILC and the systems treated in RC. But, mathematical formulations of ILC and RC are different. This monograph handles these periodic systems under the ILC framework.

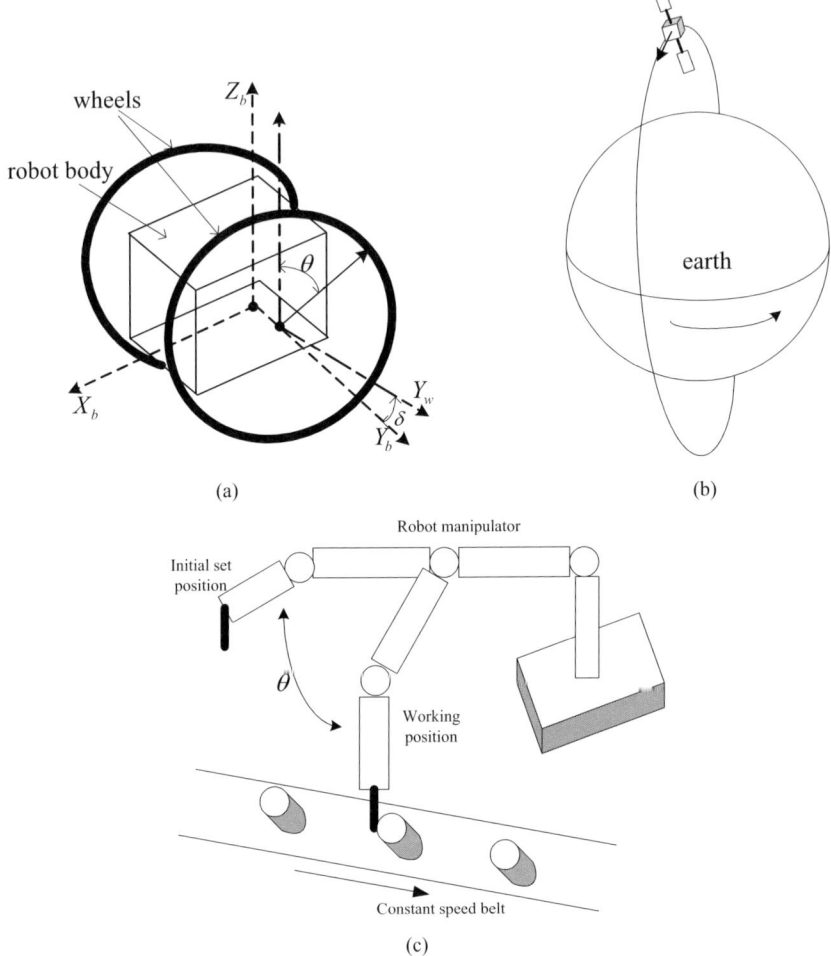

Fig. 1.2. (a): Eccentricity problem of a wheeled mobile robot. (b): Time-periodic disturbance in a satellite orbit. (c): Time-periodic robot manipulator in a manufacturing process, with an iteration-dependent disturbance.

processes, robot arm manipulators, repetitive rotary systems, and factory batch processes.
- Monotonic convergence: The system convergence to a desired trajectory can be monotonic in iteration, if convergence conditions are met. Such monotonic convergence can prevent the break-up of hardware and high overshoots of the system trajectory.
- Controller design without accurate model information: An ILC controller can be designed without an accurate model of the system. The uncertainty that can be handled by ILC includes deterministic modeling error, para-

metric uncertainty, stochastic disturbances and noise, parameter variation, and deterministic external disturbances.

1.3 Research Motivation

The principal goal of this monograph is to investigate robustness issues in ILC from an iteration-domain perspective, with an objective of demonstrating analysis and design strategies that enable monotonic convergence and perfect trajectory tracking against a variety of uncertainties. Three main motivations for this study are provided in this section.

1.3.1 Motivation for Robust Interval Iterative Learning Control

Let us consider the following single-input, single-output (SISO), 2-dimensional system described in the state-space:

$$x_k(t+1) = Ax_k(t) + Bu_k(t) \tag{1.18}$$

$$y_k(t) = Cx_k(t) \tag{1.19}$$

$$x_k(0) = x_0, \tag{1.20}$$

where $A \in \mathcal{R}^{n \times n}$, $B \in \mathcal{R}^{n \times 1}$, and $C \in \mathcal{R}^{1 \times n}$ are the matrices describing the system; $x_k(t) \in \mathcal{R}^n$, $u_k(t) \in \mathcal{R}$, and $y_k(t) \in \mathcal{R}$ are the state, input, and output vectors, respectively; t represents discrete-time points along the time axis; and the subscript k represents the iteration trial number along the iteration axis. Notice that $x_k(0) = x_0$ for all k. This is a key assumption in the ILC process and is called the initial reset condition (see again Section 1.2). Throughout this monograph, it is assumed that this initial reset condition is satisfied if there is no special indication otherwise.

For the system (1.18)–(1.20), from basic ILC theory (Theorem 1.1), a standard result is that the system is asymptotically stable (AS) if and only if $|1 - \gamma h_1| < 1$ with the ILC update law

$$u_{k+1}(t) = u_k(t) + \gamma e_k(t+1), \tag{1.21}$$

where $h_1 = CB \neq 0$ (i.e., the system has relative degree 1) and γ is the learning gain [302]. Thus, it is easy to design γ such that the condition $|1 - \gamma h_1| < 1$ is satisfied, provided that C and B are known exactly. However, generally it is reasonable to assume that there exist model uncertainties in C and B. In this case, it is necessary to select γ considering all possible model uncertainties.

In fact, this kind of ILC problem has been studied widely in the literature. For example, refer to [522, 390, 293]. However, the existing research results are almost all restricted to the asymptotic stability (AS) problem. In ILC, it has been observed that AS may not be acceptable in a practical setting, because it is possible that an ILC system can experience very high overshoots of the mean-square error during the transient before the system converges [279]. Thus, the monotonically-convergent ILC (MC-ILC) design problem has been considered to be a practically important issue, as claimed and demonstrated in [415, 279].[5] With regard to monotonically-convergent ILC, numerous publications are available. An approximate monotonic decay condition was given in [417]. A monotonic convergence condition through parameter optimization was introduced in [340]. Monotonic convergence of a Hamiltonian system was guaranteed in [140]. A maximum singular value-based monotonic convergence condition was given in [308, 331]. Following these works, in this monograph, we are thus motivated to consider not only robust asymptotic stability against interval perturbations in the system Markov parameters, but also robust monotonic convergence (i.e., exponential stability in the iteration domain).

To design a monotonically-convergent ILC algorithm, many results in the literature require information about the A matrix [298, 363, 346, 136, 9]. Though most results have considered only the nominal plant, it is again natural to consider A to be an uncertain matrix. Thus we are motivated to consider the design of the learning gain matrix when considering model uncertainties in A, B, and C.

There are a number of analysis methodologies that have been used in the ILC literature. In one such method the 2-dimensional problem of (1.18)–(1.20) is reformulated into a 1 dimensional problem. This is called the super-vector ILC (SVILC) framework [298] and it is in this framework that we will consider the robust MC-ILC problem.

To describe the SVILC methodology, take z-transforms (in time) of (1.18)–(1.20) and define the resulting plant to be $G(z)$.[6] This gives

$$
\begin{aligned}
Y(z) &= G(z)U(z) \\
&= (h_m z^{-m} + h_{m+1} z^{-(m+1)} + h_{m+2} z^{-(m+2)} + \cdots)U(z), \quad (1.22)
\end{aligned}
$$

where m is the relative degree of the system, z^{-1} is the delay operator in the discrete-time domain, and the parameters h_i are Markov parameters of the impulse response system of the plant $G(z)$. If we now define the following vectors:[7]

$$
U_k = (u_k(0), u_k(1), \ldots, u_k(N-1)) \qquad\qquad (1.23)
$$

$$
Y_k = (y_k(m), y_k(m+1), \ldots, y_k(N-1+m)) \qquad\qquad (1.24)
$$

$$
Y_d = (y_d(m), y_d(m+1), \ldots, y_d(N-1+m)) \qquad\qquad (1.25)
$$

$$
E_k = Y_d - Y_k = (E_k(m), E_k(m+1), \ldots, E_k(N-1+m)), \qquad (1.26)
$$

then the linear plant can be described by $Y_k = HU_k$, where H is a Toeplitz matrix of rank N whose elements are the Markov parameters of the plant

---

[5] For more precise definitions of AS and MC, see Section 3.2 and Section 4.1.
[6] $G(z) = C(zI - A)^{-1}B + D$.
[7] The process of forming "super-vectors" U_k, Y_k, Y_d, and E_k is called "lifting" in the literature [298].

$G(z)$, given by:

$$
H = \begin{bmatrix}
h_m & 0 & 0 & \cdots & 0 \\
h_{m+1} & h_m & 0 & \cdots & 0 \\
h_{m+2} & h_{m+1} & h_m & \cdots & 0 \\
\vdots & \vdots & \vdots & \ddots & \vdots \\
h_{m+N-1} & h_{m+N-2} & h_{m+N-3} & \cdots & h_m
\end{bmatrix}. \tag{1.27}
$$

Throughout this monograph, this matrix is called the system Markov matrix or simply the Markov matrix.

The monotonic convergence condition for the system (1.18)–(1.20) with the standard Arimoto-type ILC update law (1.21) is now simply given as

$$
\|I - H\Gamma\| < 1,
$$

where $\|\cdot\|$ is an operator norm and Γ is the learning gain matrix. Γ is a diagonal matrix whose diagonal elements are the scalar learning gains γ.

In the super-vector framework we just described, the system given in (1.18)–(1.20) is assumed to be the nominal plant, i.e., neither model uncertainty nor process disturbances or measurement noises are considered. Thus, the MC condition $\|I - H\Gamma\| < 1$ is not practically meaningful when taking into account model uncertainties or external disturbances and noise. To address this, consider instead the following 2-dimensional uncertain plant model:

$$
x_k(t+1) = (A + \Delta A)x_k(t) + (B + \Delta B)u_k(t) + v(k,t) \tag{1.28}
$$
$$
y_k(t) = (C + \Delta C)x_k(t) + w(k,t), \tag{1.29}
$$

where ΔA, ΔB, and ΔC are model uncertainties, and $v(k,t)$ and $w(k,t)$ are time- and iteration-dependent process disturbance and measurement noise signals, respectively. Recall that t is the discrete-time point along the time axis, which means that t is defined in a finite interval. That is, on each iteration or trial, t has N different discrete-time points. Meanwhile, k is defined on an infinite horizon, so it increases monotonically. Thus we can consider two different types of model uncertainties. The first type is iteration-independent model uncertainty, while the second type is iteration-dependent model uncertainty.

We are interested in addressing the robustness and convergence properties of systems such as (1.28)–(1.29). However, the SVILC framework requires analysis based on the system Markov matrix. Thus, it becomes necessary to convert the model uncertainty of (1.28)–(1.29) to uncertainty associated with the Markov matrix (1.27). To our knowledge, this uncertainty conversion problem has never been addressed in the existing literature except in our own work. This problem, which we call "interval model conversion," will be carefully addressed in this monograph. In particular, for the interval model conversion problem, the power of an interval matrix will need to be computed and we will show how this can be done in a computationally efficient and non-conservative fashion.

Next, let us suppose that Markov matrix includes the model uncertainty converted from the uncertain plant (1.28)–(1.29) and let us denote the uncertain system Markov matrix as $H = H^o + \Delta H$, where H^o is the system Markov matrix corresponding to the nominal plant and ΔH is the uncertain system Markov matrix corresponding to the uncertainty of the uncertain plant. Then, our task is to find a learning gain matrix such that the system is AS or MC against all possible uncertain plants H. There are two issues with regard to this robust ILC design problem. The first issue is related to the conservativeness and the computational cost of the proposed method, and the second issue is related to the performance of the algorithm. The performance issue is concerned with the stability types: asymptotic stability (AS) and monotonic convergence (MC).[8] As noted, although there are numerous results on asymptotic and robust stability in ILC [522, 390, 183], to date there is no systematic analysis and synthesis framework addressing robust monotonically-convergent ILC. Most existing works have focused on asymptotic stability for plants with model uncertainty described in the state-space form given by (1.28)–(1.29). To our knowledge, outside of our own work, no systematic approach for handling the uncertainty in H (i.e., ΔH) has been reported. Furthermore, though in many existing works, optimal-ILC [9, 181, 342, 341], stochastic noise [397, 396, 395, 54], and frequency-dependent uncertainty [360] have been considered, these techniques can give conservative results. We show, however, that using parametric interval concepts can reduce the conservatism connected with robustness tests and can provide tighter monotonic convergence conditions.

In summary, there are three main research motivations for studying the robust interval ILC problem. The first motivation is to solve fundamental problems associated with the robustness of ILC designs when the plant is subject to parametric uncertainty. Then, based on these results, the second task is to guarantee monotonic convergence against all possible interval uncertainties. Finally, the goal is to reduce the conservatism related to robust stability tests.

1.3.2 Motivation for H_∞ Iterative Learning Control

From the ILC literature, there is no systematic approach to handling iteration-varying model uncertainty, iteration-varying external disturbances, or iteration-varying stochastic noise all together [360]. Even though time-domain-based

---

[8] Note that stability in an ILC problem refers to the boundedness of signals at fixed points of time, considered along the iteration axis. By assumption (due to the finite-time horizon), traditional stability along the time axis is achieved by default (except, perhaps, in the case of a nonlinear system exhibiting finite escape-time behaviors). Thus, by AS, we mean that the ILC system converges to the desired trajectory as the iteration number increases. By MC, we mean that the ILC system converges without overshoot to the desired trajectory as iteration number increases.

H_∞ ILC schemes and 2-dimensional approaches have been suggested for making a unified ILC framework, research to date has been limited to iteration-independent uncertainty and disturbances. However, if we can cast the super-vector notation as defined by (1.23)–(1.26) into a traditional discrete-time H_∞ framework (where now "time" is actually "iteration"), we can obtain a unified robust control framework on the iteration domain. Furthermore, our proposed unified robust H_∞ ILC approach on the iteration domain provides a way to consider and discuss frequency-domain analysis on the iteration axis. Recall that in ILC the time axis is finite-horizon. Thus, the corresponding frequency domain transformed from the time axis should also be finite. However, in control engineering the frequency domain is usually infinite. Thus, the frequency-domain-based ILC analysis, transformed from the finite time axis, is not suitable from an analytical perspective. However, in our proposed H_∞ scheme on the iteration axis, the discrete infinite iteration axis is readily transformed into the discrete infinite frequency domain. Hence, the H_∞ ILC scheme on the iteration axis provides a unified robust control framework on the infinite frequency domain that is analytically correct.

1.3.3 Motivation for Stochastic Iterative Learning Control

Even though the H_∞ ILC scheme on the iteration domain presented in this monograph provides a unified framework, it is not related to monotonic convergence. Motivated by this observation, we can raise a question: is it possible to design a learning gain matrix to ensure monotonic convergence when considering stochastic noise or iteration-varying model uncertainty? Further, a related question is how to analytically estimate the steady-state error that the ILC system can achieve under stochastic noise and model uncertainty. In the stochastic ILC approach proposed in this monograph, we try to estimate the ultimate baseline error of an uncertain ILC system over which monotonic convergence is guaranteed. And, we wish to make this estimate in an off-line manner. A related problem also arises in the area of networked-control systems (NCS), where data dropout problems have been popularly studied. In this monograph, we try to integrate the NCS into an ILC framework so that an overall intermittent ILC system can be developed that is robust against extreme data dropout situations in a network.

1.4 Original Contributions of the Monograph

This monograph makes the following theoretical contributions:

- Conditions for robust stability in the iteration domain are provided for parametric interval systems.
- Techniques for converting from time-domain interval models to Markov interval models are given.

- A monotonically-convergent ILC system is designed under parametric interval uncertainties and/or stochastic noise considerations.
- Robust H_∞ ILC is designed on the iteration domain, taking into account three different types of uncertainties: iteration-variant/invariant model uncertainty, external disturbances, and stochastic noise.
- The baseline error of the ILC process is analytically established, which provides a novel idea for designing the ILC learning gain matrix in an off-line manner.
- Solutions for three fundamental interval computational problems are offered: robust stability of an interval polynomial matrix system, the power of an interval matrix, and the maximum singular value of an interval matrix.

It is the main contribution of this monograph to provide new analytical tools for designing robust ILC systems. Using the super-vector approach, the robustness problem of ILC is discussed purely on the iteration domain. Parametric interval uncertainty enables us to design both the monotonically-convergent ILC process and a less conservative robust ILC system. Indeed, this robust, monotonically-convergent ILC design method makes a significant contribution to practical ILC applications because it avoids unacceptable overshoot on the iteration domain while considering all possible models the controller might face. Furthermore, by casting the H_∞ framework and Kalman filtering into the SVILC framework, the monograph provides a different design perspective for stochastic and frequency-domain uncertain ILC systems than has typically been found in the literature. Additionally, analytical solutions for the three fundamental interval computational problems mentioned above are provided. These solutions can be effectively used for solving various types of control and systems problems. For example, robust controllability, robust observability, multi-input multi-output robust control theory, robust monotonically-convergent stability problems, and robust boundary calculations for the model conversion problem can all be addressed using the results presented in the monograph.

2

An Overview of the ILC Literature

Historically, the ILC idea appeared perhaps as early as 1970 in a U.S. patent, as explained in [59]. However, in the scientific literature the first idea related to ILC was possibly the multipass control concept, which can be traced back to Edwards in 1974 [125], though the stability analysis was restricted to classical control concepts and an explicit ILC formulation was not given. It is widely accepted that the initial explicit formulation of ILC was given by Uchiyama, in Japanese, in 1978 [461] and by Arimoto, in English, in 1984 [28].

In the published literature, major ILC surveys were performed in 1992, 1997, and 1999, in Section 2 of [307], Chapter 1 of [68], and Chapter 4.4 of [299], respectively. The early part of Section 2 of [307] introduced Japanese researchers who suggested LTI Arimoto-type gains, PID-type gains, and gradient method-based optimization algorithms. Then, in the latter part of Section 2 of [307], literature dealing with nonlinear ILC, robustness of ILC, adaptive schemes in ILC, optimal ILC strategies, and neural network-based approaches were introduced. For more detailed explanations about the early research before 1990, see Section 2 of [307]. The first classification of ILC research was given in Chapter 1 of [68] and a wider ILC classification was performed in Chapter 4.4 of [299]. Notably, in [299], a literature review of work before 1997 was performed based on two categories: theoretical works in ILC and ILC applications. In total, 256 publications were covered in [299], which were identified using a search on keywords "control" AND "learning" AND "iterative." For a more recent survey, see [47], which provides a detailed technical survey of ILC algorithms along with new results on the design of the so-called Q-filter.

In the remainder of this chapter, we present a summary of the ILC literature from 1998 to 2004. We begin by describing the methodology and numerical results of a detailed literature search. We then give a number of comments on the results of our literature search, with specific emphasis on a special issue of the *International Journal of Control* and on a number of Ph.D. dissertations on ILC. Finally, we refer the reader to Appendix A, where we have included a taxonomy that categorizes the ILC literature between 1998 and 2004 into two different parts, following [299]. The first part is related to the publications that focus on ILC applications and the second part is related to the publications that handle theoretical developments.

2.1 ILC Literature Search Methodology

Table 2.1 shows the results from searching the "Web of Science"[1] and the "IEEE Xplore"[2] databases on January 4, 2005. As shown in this table, from the keywords "control" AND "learning" AND "iterative," we have a total of 877 publications. Broader searches were executed using "Iterative" AND "Learning" and "Learning" AND "Control," from which 1910 and 20, 260 publications were found, respectively. Hence, associated with the keyword "Learning," a great amount of research has been published. Associated with "Repetitive Control," we see that 309 publications were published. It is almost impossible to systematically review all these publications at this point, so in this monograph the literature review is restricted to the publications found by searching on the exact phrase "Iterative Learning Control," from which 510 different publications were found. However, for a reliable taxonomy, conference papers and journal papers that cannot be searched in the Web of Science and the IEEE Xplore database are also included. They are taken from the 1999 and 2002 International Federation of Automatic Control (IFAC) World Congress (WC) papers, the 2000 and 2002 Asian Control Conference (ASCC) papers, *Asian Journal of Control* (AJC) papers, the 2001 European Control Conference (ECC) papers, the 7th Mechatronics Conference papers, and a variety of miscellaneous conference papers (see Table 2.2). Thus, our review covers IEEE conferences and journal papers, international (SCI) journal papers, IFAC Conference papers, and Asian Control Conference papers. Figure 2.1 shows the number of publications since 1990 in these international conference proceedings and journals. As shown in this plot, the number of publications grew steadily between 1990 and 1998 and then grew very quickly through 2002. A drop off in publications is shown in 2003 and 2004. Table 2.3 shows the regional distribution of the authors of the publications that were published in IEEE Conference proceedings and SCI journals (from Web of Science).

2.2 Comments on the ILC Literature

The first ILC monograph [298] was published in 1993. There was an edited book [44] in 1998; subsequently there were three special issues of journals (a special issue of the *International Journal of Control* [312] in 2000, a special

---

[1] http://isi01.isiknowledge.com/portal.cgi/wos.
[2] http://ieeexplore.ieee.org.

Table 2.1. ILC-related Publications From Web of Science and IEEE Xplore

Search options	Web of Science	IEEE Conference	Total
Iterative + Learning	793	1117	1910
Learning + Control	12, 739	7521	20, 260
Iterative + Learning + Control	367	510	877
Iterative Learning Control	241	269	510
Repetitive Control	150	159	309

Table 2.2. Miscellaneous ILC-related Publications from 1998 to 2004

IFAC 1999 WC	IFAC 2002 WC	2000 ASCC	2002 ASCC	AJC	Others	Total
12	19	20	11	14	7	83

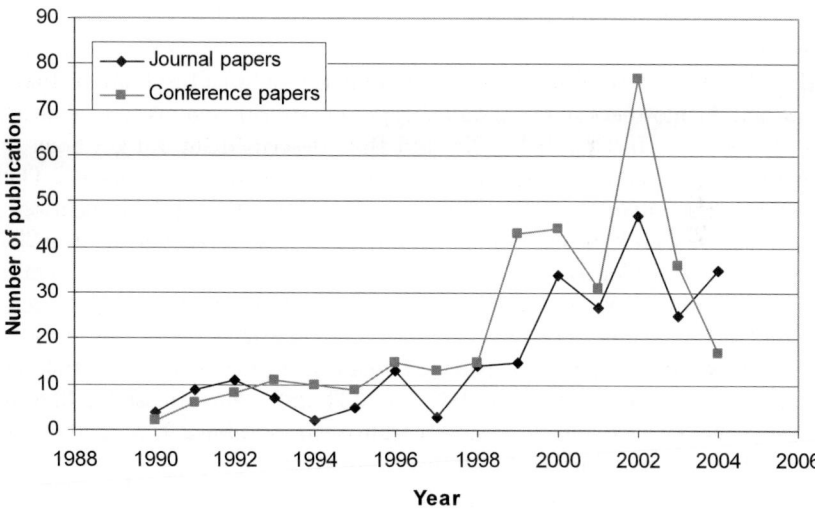

Fig. 2.1. Publication numbers of ILC-related literature in conference proceedings and journals

issue of the *Asian Journal of Control* in 2002, and a special issue of the journal *Intelligent Automation and Soft Computing* [40]); two more ILC monographs [66, 518] appeared in 1999 and in 2003, respectively. We would note that [44] is the outcome of several invited sessions at the 2nd Asian Control Conference held in Seoul, Korea in July, 1997 and [312] is the outcome of a roundtable on ILC held in conjunction with the 1998 IEEE Conference on Decision and Control in Tampa, Florida, USA.

Table 2.3. Regional Distribution of Authors of IEEE Conference Papers and SCI Journal Papers

Region	From "Web of Science"	From "IEEE Conference"	Total
China	18	55	73
Korea	60	15	75
Singapore	68	39	107
Japan	32	30	62
Taiwan	9	5	14
UK	47	29	76
Europe	42	30	72
USA	63	39	102
Canada	11	8	19
Other areas	6	6	12

It is useful to read Chapter 1 and Chapter 2 of [44]. In Chapter 1, as a conclusion, Arimoto argued that the P-type update rule may be more natural than D-type ILC. In Chapter 2, Xu and Bien described several key issues in ILC research and commented on the limitations of ILC applications. Their discussions were given in three different categories: tasks, the connection of ILC to other control theories, and future research directions in ILC. In [66] nonlinear higher-order ILC was developed to address robust ILC stability and in [518] nonlinear ILC, mostly based on the idea of a composite-energy function, was described.

Since 1998, at least 18 different Ph.D. dissertations can be found [184, 332, 467, 114, 538, 210, 159, 137, 554, 287, 540, 334, 282, 484, 416, 263, 276, 367]. These dissertations, which are enumerated by year in Table 2.4, were identified by searching the "Digital Dissertations" web-site.[3]

Table 2.4. ILC-related Ph.D. Dissertations

	1998	1999	2000	2001	2002	2003	2004	Total
Number	1	1	3	2	3	0	8	18

In the next two sections we present a review and comments on the special issue on ILC of the *International Journal of Control* (IJC), Vol 73, No 10, [312], which we consider to contain a nice snapshot of ILC at the turn of the century, and we comment on the Ph.D. dissertations noted above.

---

[3] http://wwwlib.umi.com/dissertations/gateway.

2.3 IJC Special Issue

In [30], Arimoto discussed the equivalence between the concepts of "learnability," "output-dissipativity," and "strictly positive realness." Using Theorems 1–4 given in [30], it is possible to check if there exists an ILC controller that ensures the input–output $\mathcal{L}_2$ stability of the controlled system. For instance, in the case where $D = 0$ in the state-space description, learnability can be checked by investigating if there exist two positive definite symmetric matrices X and Q such that

$$A^T X + XA = -Q, \quad XB = C^T. \tag{2.1}$$

This approach has several potential applications in ILC research, because the Lyapunov-like equation given in (2.1) can be used in various ways as done in other control areas. In [136], a linear quadratic ILC scheme was modified for practical applications. One of the main novelties of [136] is the dimensional reduction of the super-vectors in calculating the optimal control at each trial. Another novelty is that the unknown system model was estimated based on conjugate basis vectors. French and Rogers [134] provided an adaptive ILC scheme with a calculated cost for the $\mathcal{L}_p$ bounded disturbance. It is the main achievement of [134] to handle the robustness issue in an adaptive control framework. Owens and Munde [343] also provided a new adaptive approach for ILC. They included current cycle error feedback in the adaptive control law. This is motivated by the fact that the most recent error data reflects the current performance most closely. Also, by including the current cycle feedback signal, they can stabilize unstable plants during each trial. In [522], Xu et al. suggested a robust learning controller (RLC) for robotic manipulators to compensate for state-independent periodic uncertainties and to suppress nonperiodic system uncertainties. As commented in the same paper, the results of [522] can be applied to various periodically disturbed systems and uncertain dynamic systems. In [356], the initial state error problem was addressed, but the robustness issue was not fully answered. In [196], Hillenbrand and Pandit provided a design scheme with reduced sampling rate. A concept called anticipatory ILC was suggested by Wang in [474], whereby the ILC update rule is given by

$$u_{i+1}(t) = u_i(t) + L(\cdot)[y_d(t + \triangle) - y_i(t + \triangle)] \tag{2.2}$$

with a saturation condition. It is interesting to note that the update rule is different from D-type or P-type. In [82], Chien suggested an ILC design method based on a fuzzy network for sampled-data systems and in [391], the state observer and disturbance model were used in the learning controller. A particularly valuable paper in the IJC 2000 special issue was presented by Longman [279], who provided several important guides for actual application of ILC and repetitive control (RC) algorithms. Longman also provided experimental test results and detailed explanations about the tests for the practical uses of ILC. In [309], Moore proved the convergence of ILC algorithms to achieve a desired periodic output trajectory. As extensions of the results of [309], periodic disturbance compensation and practical applications were also discussed. The last three papers of [312] were dedicated to ILC applications. Specifically, [390] used an H_∞ technique for wafer positioning control, [337] applied ILC to non-holonomic systems, and [38] showed that ILC can be utilized for position control of chain conveyer systems.

2.4 ILC-related Ph.D. Dissertations Since 1998

We begin this section by noting that our search for Ph.D. dissertations published since 1998 is very limited, because "Digital Dissertations" does not include all the schools in the world and we were not able to be personally aware of all the dissertations published everywhere on this topic. Nonetheless, we tried to include all the Ph.D. dissertations of which we were aware. In 2004, the number of Ph.D. dissertations in ILC significantly increased as shown in Table 2.4. In [184], Hätönen studied the algebraic properties of a standard ILC structure and made progress on the topic of norm-optimal ILC. Verwoerd [467] suggested equivalent feedback controllers for causal ILC and noncausal ILC based on an admissibility concept. Note that a similar discussion to [467] can be found in [162, 345, 163]. Dijkstra [114] showed some exciting ILC applications. In his dissertation, lower-order ILC was applied to the different wafer-stages. In addition, for finite-time ILC, Dijkstra provided several interesting theoretical developments in Chapter 4 of [114]. In [334] Oh introduced a local learning concept to avoid undesirable overshoot during the learning transient. Norrlöf [332] presented a number of useful results on the theory of ILC, including ideas about the use of models in ILC and presentation of a successful ILC application for a robotic manipulator. Markusson [287] used ILC to find an inversion of the system, particulary focused on noncausal and nonminimum systems. A time–frequency adaptive Q-filter ILC was suggested for nonsmooth-nonlinearity compensation by Zheng in [554] and the idea was used for an injection molding machine. ILC and RC were summarized and some new results for nonlinear, nonminimum phase systems were developed by Ghosh in [159]. In [538], Yang studied ILC based on neural networks and in [137], Frueh suggested a basis-function-model-reference adaptive learning controller (see also [136]). The suggested method in [137] has several advantages. One in particular is an adaptive property to account for slowly varying plant parameters or errors from the initial model. Huang [210] introduced Fourier series-based ILC algorithms for tracking performance improvement. In [484], several important issues in the field of learning and repetitive control were addressed by Wen, including indirect adaptive control ideas applied to learning control and basis-functions used to show that ILC and RC problems are mathematically the same under certain conditions. Songchon showed, in [416] that learning control has the ability to bypass the waterbed effect, which is a fundamental problem in traditional feedback controls. In [263], LeVoci developed methods for predicting the final error levels of general first-order ILC, of higher-order ILC with current cycle learning, and of general RC, in the presence of noise, using frequency response methods. Three main difficulties in the area of linear discrete-time ILC were addressed by Lo in [276]: (i) The number of output variables for which zero tracking error can be achieved is limited by the number of input variables; (ii) every variable for which zero tracking error is sought must be a measured variable; and (iii) in a digital environment, the inter-sample behavior may have undesirable error from ripple. An interesting application of optimal ILC was utilized for a chemical molding process by Yang in [540]; Ma [282] showed that an ILC algorithm can be used for vision-based tracking systems; and in [367], Phetkong used ILC on a cam designed and built using a polynomial profile and it was shown that eight cycles of learning were sufficient to effectively accomplish the morphing of the cam behavior.

2.5 Chapter Summary

In this chapter, we provided a brief overview of the ILC literature, including comments on selected results. We also pointed the reader to a complete taxonomy of the literature on ILC between 1998 and 2004 that is included in Appendix A. From this taxonomy we can understand the overall trend of ILC research from both the application point of view and the theoretical point of view.

3

The Super-vector Approach

This chapter is devoted to a technical overview of the super-vector ILC (SVILC) framework. The asymptotic stability and monotonic convergence of first-order SVILC systems have been well-described in [331]. Thus, here we focus on higher-order SVILC systems. Specifically, asymptotic stability (AS) and monotonic convergence (MC) conditions for higher-order ILC (HOILC) system are studied.

Throughout this chapter the discrete-time plant given in (1.18)–(1.20), with $G(z) = C(zI - A)^{-1}B + D$, is considered. It is assumed that $t \in [0, T]$. In Chapter 1, the Markov matrix H was given in (1.27). Without loss of generality, we take $m = 1$ and $CB \neq 0$ in the expression for H. The following HOILC update rule is considered [301]:

$$U_{k+1} = \Lambda_k U_k + \Lambda_{k-1} U_{k-1} + \cdots + \Lambda_{k-n} U_{k-n}$$
$$+ \Gamma_{k+1} E_{k+1} + \Gamma_k E_k + \cdots + \Gamma_{k-n} E_{k-n}, \tag{3.1}$$

where k denotes the iteration trial; $\Lambda_i, i = k - n, \ldots, k$, are fixed learning gain matrices from the previous control input vectors; n represents the number of the past trials used for the current control update (if $n = 0$, it is first-order ILC, if $n \geq 1$, it is HOILC); and $\Gamma_i, i = k - n, \ldots, k$ are fixed learning gain matrices from the previous error vectors.

3.1 Asymptotic Stability of Higher-order SVILC

In this section, we summarize some results of [3]. From the definition $Y_k = Y_d - E_k$, with the assumption $h_1 \neq 0$ (note: if $h_1 \neq 0$, then H is nonsingular), and after changing $Y_k = H U_k$ to $U_k = H^{-1} Y_k$, we substitute $U_k = H^{-1} Y_k$ into (3.1). Then, after several algebraic simplifications (refer to [3]), we obtain the following relationship:

$$H^{-1} Y_{k+1} = \Lambda_k H^{-1} Y_k + \cdots + \Lambda_{k-n} H^{-1} Y_{k-n}$$
$$+ \Gamma_{k+1} E_{k+1} + \Gamma_k E_k + \cdots + \Gamma_{k-n} E_{k-n}. \tag{3.2}$$

By inserting $Y_k = Y_d - E_k$ into (3.2), we have

$$
\begin{aligned}
H^{-1}(Y_d - E_{k+1}) &= \Lambda_k H^{-1}(Y_d - E_k) + \cdots + \Lambda_{k-n} H^{-1}(Y_d - E_{k-n}) \\
&\quad + \Gamma_{k+1} E_{k+1} + \Gamma_k E_k + \cdots + \Gamma_{k-n} E_{k-n} \\
&= (\Lambda_k + \cdots + \Lambda_{k-n}) H^{-1} Y_d - \Lambda_k H^{-1} E_k - \\
&\quad \cdots - \Lambda_{k-n} H^{-1} E_{k-n} + \Gamma_{k+1} E_{k+1} + \\
&\quad \cdots + \Gamma_{k-n} E_{k-n}.
\end{aligned}
\tag{3.3}
$$

A constraint $\Lambda_k + \cdots + \Lambda_{k-n} = I_{N \times N}$ is used to make the steady-state error zero (from now on, I represents $I_{N \times N}$). Then (3.3) is changed to

$$
\begin{aligned}
(\Gamma_{k+1} + H^{-1}) E_{k+1} &= (\Lambda_k H^{-1} - \Gamma_k) E_k + \cdots \\
&\quad + (\Lambda_{k-n} H^{-1} - \Gamma_{k-n}) E_{k-n}.
\end{aligned}
\tag{3.4}
$$

For convenience, writing $\mathcal{A} := \Gamma_{k+1} + H^{-1}$ and $\mathcal{B}_i := \Lambda_i H^{-1} - \Gamma_i$, (3.4) can be rewritten as

$$
\mathcal{A} E_{k+1} = \mathcal{B}_k E_k + \cdots + \mathcal{B}_{k-n} E_{k-n}.
\tag{3.5}
$$

Thus, if $\mathcal{A}$ is nonsingular (which can always be assured by choice of Γ_{k+1}; of particular importance below, when $\Gamma_{k+1} = 0$ then $\mathcal{A} - H^{-1}$ is in fact nonsingular), then

$$
E_{k+1} = \mathcal{A}^{-1} \mathcal{B}_k E_k + \cdots + \mathcal{A}^{-1} \mathcal{B}_{k-n} E_{k-n}.
\tag{3.6}
$$

Now, taking the w-transform along the iteration axis,[1] (3.6) is changed to

$$
\begin{aligned}
&E(w) w^{k+1} - \mathcal{A}^{-1} \mathcal{B}_k E(w) w^k - \cdots - \mathcal{A}^{-1} \mathcal{B}_{k-n} E(w) w^{k-n} = 0 \\
&\Leftrightarrow E(w)\left(w^{k+1} - \mathcal{A}^{-1} \mathcal{B}_k w^k - \cdots - \mathcal{A}^{-1} \mathcal{B}_{k-n} w^{k-n}\right) = 0.
\end{aligned}
\tag{3.7}
$$

Hence, using the notation

$$
P(w) := w^{k+1} I - \mathcal{A}^{-1} \mathcal{B}_k w^k - \cdots - \mathcal{A}^{-1} \mathcal{B}_{k-n} w^{k-n},
\tag{3.8}
$$

we can say the HOILC system is AS if and only if $P(w)$ is stable. The stability of the polynomial matrix $P(w)$ is defined later in Definition C.1.

Remark 3.1. Actually, it is a well-known fact [13] that the stability of $P(w)$ can be checked by the companion matrix form given by

$$
C_{P(w)} := \begin{bmatrix}
\mathcal{A}^{-1} \mathcal{B}_k & \mathcal{A}^{-1} \mathcal{B}_{k-1} & \mathcal{A}^{-1} \mathcal{B}_{k-2} & \cdots & \mathcal{A}^{-1} \mathcal{B}_{k-n} \\
I & 0 & 0 & \cdots & 0 \\
0 & I & 0 & \cdots & 0 \\
\vdots & \vdots & \vdots & \ddots & \vdots \\
0 & 0 & 0 & \cdots & 0
\end{bmatrix}.
$$

If the spectral radius of $C_{P(w)}$ is less than 1, then $P(w)$ is considered stable.

---

[1] In [301], the w-domain was defined by $w^{-1} u_k(t) = u_{k-1}(t)$, where k is the iteration trial number and t is the discrete-time point. This will be described in more detail later in the monograph.

3.2 Monotonic Convergence of Higher-order SVILC

In the literature, analytical monotonic convergence conditions for higher-order SVILC have not been reported. In this section, we derive such conditions. As a conclusion, it will be shown that it is difficult to guarantee the MC of higher-order ILC systems.

Let us first briefly consider first-order ILC (FOILC). For this case, referring back to (3.1), we will assume that $\Gamma_{k+1} = 0$.[2] Then, the ILC process is called monotonically-convergent (MC) if $\|E_{k+1}\| < \|E_k\|$, and the ILC process is called semi monotonically convergent (SMC) if $\|E_{k+1}\| \leq \|E_k\|$ (for the formal definition, see Definition 4.1 in Chapter 4). From (3.4), the error evolution rule is expressed in the FOILC case, with $\Gamma_{k+1} = 0$, as

$$E_{k+1} = (I - H\Gamma_k)E_k. \tag{3.9}$$

Then, from the following relationship:

$$\|E_{k+1}\| = \|(I - H\Gamma_k)E_k\| \leq \|I - H\Gamma_k\|\|E_k\|, \tag{3.10}$$

the MC condition of FOILC is derived as $\|I - H\Gamma_k\| < 1$ and the SMC condition is derived as $\|I - H\Gamma_k\| \leq 1$.

Next, for HOILC, the following definition is suggested:

Definition 3.2. *In HOILC, the ILC process is monotonically-convergent (MC) in an appropriate norm topology if*

$$\|E_{k+1}\| < \max\{\|E_i\|, \ i = k, \ldots, k - n\}, \tag{3.11}$$

and the ILC process is semi monotonically convergent if

$$\|E_{k+1}\| \leq \max\{\|E_i\|, \ i = k, \ldots, k - n\}. \tag{3.12}$$

∎

We may now state the following result:

Theorem 3.3. *In HOILC, if $\mathcal{A}$ is nonsingular and $\underline{\sigma}(\mathcal{A}) > \sum_{i=k-n}^{i=k} \|\mathcal{B}_i\|_2$, then the ILC process is MC in the l_2-norm topology.* ∎

Proof. From (3.5), using the matrix 2-norm, we have

$$
\begin{aligned}
\|E_{k+1}\| &= \|\mathcal{A}^{-1}\mathcal{B}_k E_k + \cdots + \mathcal{A}^{-1}\mathcal{B}_{k-n}E_{k-n}\| \\
&\leq \|\mathcal{A}^{-1}\|\|\mathcal{B}_k\|\|E_k\| + \cdots + \|\mathcal{A}^{-1}\|\|\mathcal{B}_{k-n}\|\|E_{k-n}\| \\
&= \|\mathcal{A}^{-1}\|(\|\mathcal{B}_k\|\|E_k\| + \cdots + \|\mathcal{B}_{k-n}\|\|E_{k-n}\|) \\
&\leq \|\mathcal{A}^{-1}\|(\|\mathcal{B}_k\| + \cdots + \|\mathcal{B}_{k-n}\|)\max\{\|E_i\|\} \\
&= \frac{1}{\underline{\sigma}(\mathcal{A})}(\|\mathcal{B}_k\| + \cdots + \|\mathcal{B}_{k-n}\|)\max\{\|E_i\|\},
\end{aligned} \tag{3.13}
$$

---

[2] Note that this means there is no current-cycle (or current-iteration) feedback (CITE).

where $\| \cdot \|$ is the matrix 2-norm. Therefore, if $\underline{\sigma}(\mathcal{A}) > \sum_{i=k-n}^{i=k} \|\mathcal{B}_i\|$, then $\|E_{k+1}\| < \max\{\|E_i\|, \ i = k, \ldots, k - n\}$. ∎

The previous theorem is quite general. If we let $\Gamma_{k+1} = 0$ (i.e., there is no current-cycle feedback), then we can give a more specific result as follows:

Theorem 3.4. *If H is nonsingular and $\Gamma_{k+1} = 0$, then a sufficient condition for monotonic convergence is $\sum_{i=k-n}^{i=k} \|H\Lambda_i H^{-1} - H\Gamma_i\| < 1$.* ∎

Proof. From (3.4),

$$H^{-1}E_{k+1} = (\Lambda_k H^{-1} - \Gamma_k)E_k + \cdots + (\Lambda_{k-n} H^{-1} - \Gamma_{k-n})E_{k-n}$$
$$\Leftrightarrow E_{k+1} = (H\Lambda_k H^{-1} - H\Gamma_k)E_k + \cdots + (H\Lambda_{k-n}H^{-1} - H\Gamma_{k-n})E_{k-n}. \tag{3.14}$$

Let us take a matrix norm on both sides to get:

$$
\begin{aligned}
\|E_{k+1}\| &= \|(H\Lambda_k H^{-1} - H\Gamma_k)E_k + \cdots + (H\Lambda_{k-n}H^{-1} - H\Gamma_{k-n})E_{k-n}\| \\
&\leq \|H\Lambda_k H^{-1} - H\Gamma_k\|\|E_k\| + \cdots + \|H\Lambda_{k-n}H^{-1} - H\Gamma_{k-n}\|\|E_{k-n}\| \\
&\leq \sum_{i=k-n}^{i=k} \|H\Lambda_i H^{-1} - H\Gamma_i\| \max\{\|E_i\|\}.
\end{aligned}
\tag{3.15}
$$

Therefore, if $\sum_{i=k-n}^{i=k} \|H\Lambda_i H^{-1} - H\Gamma_i\| < 1$, then $\|E_{k+1}\| < \max\{\|E_i\|, \ i = k - n, \ldots, k\}$. This completes the proof. ∎

Unfortunately, the condition of Theorem 3.4 is quite conservative for checking the monotonic convergence of the HOILC system, because it is difficult to achieve the condition $\sum_{i=k-n}^{i=k} \|H\Lambda_i H^{-1} - H\Gamma_i\| < 1$. The argument is as follows. If, for example, $\Lambda_i, i = k - n, \ldots, k$ are lower-triangular Toeplitz matrices, then from the following relationship:

$$\sum_{i=k-n}^{i=k} \|\Lambda_i - H\Gamma_i\| \geq \left\| \sum_{i=k-n}^{i=k} (\Lambda_i - H\Gamma_i) \right\| \tag{3.16}$$

and using $\sum_{i=k-n}^{i=k} \Lambda_i = I$, we have

$$
\begin{aligned}
\left\| \sum_{i=k-n}^{i=k} (\Lambda_i - H\Gamma_i) \right\| &= \left\| I - \sum_{i=k-n}^{i=k} H\Gamma_i \right\| \\
&= \left\| I - H\left(\sum_{i=k-n}^{i=k} \Gamma_i \right) \right\|.
\end{aligned}
\tag{3.17}
$$

Therefore, if $\|I - H\Gamma_k\| > 1$ in FOILC and $\Gamma_i, \ i < k$, have the same structure as Γ_k, then $\sum_{i=k-n}^{i=k} \|\Lambda_i - H\Gamma_i\|$ is always bigger than 1 in the HOILC case, because $\left\| I - H\left(\sum_{i=k-n}^{i=k} \Gamma_i \right) \right\| > 1$.

Thus, to reduce the conservatism, let us define another monotonic convergence concept and condition.

Definition 3.5. *If* $\max\{\|E_i\|, i = k + 1, \ldots, k - n + 1\} < \max\{\|E_i\|, i = k,$ $\ldots, k - n\}$, *then the ILC process is called block-wise monotonically-convergent (BMC). Similarly, if*

$$\max\{\|E_i\|, i = k + 1, \ldots, k - n + 1\} \leq \max\{\|E_i\|, i = k, \ldots, k - n\}, \quad (3.18)$$

then the ILC process is called block-wise semi-monotonically-convergent (BSMC). ∎

To proceed we also need to introduce the following definition:

Definition 3.6. *The l_1-norm of the pth row vector of the matrix A is denoted $\|A_p\|$. For example, $\|A_1\|$ represents the l_1-norm of the first row vector of the matrix A.* ∎

We may now state the following result:

Theorem 3.7. *If $\Gamma_{k+1} = 0$ and Λ_i are lower-triangular Toeplitz matrices, then the ILC process is BSMC in the l_1-norm topology if*

$$\max\left\{\sum_{i=k-n}^{i=k} \|(\Lambda_i - H\Gamma_i)_p\|_1, \ p = 1, \ldots, N\right\} < 1,$$

where N is the size of the square Markov matrix H. ∎

Proof. To prove Theorem 3.7, we use the matrix companion form associated with

$$E_{k+1} = C_k E_k + \cdots + C_{k-n} E_{k-n}, \quad (3.19)$$

where $C_i := (\Lambda_i - H\Gamma_i)$. The companion form of (3.19) is expressed as

$$
\begin{bmatrix}
E_{k+1} \\
E_k \\
E_{k-1} \\
\vdots \\
E_{k-n+1}
\end{bmatrix}
=
\begin{bmatrix}
C_k & C_{k-1} & C_{k-2} & \cdots & C_{k-n} \\
I & 0 & 0 & \cdots & 0 \\
0 & I & 0 & \cdots & 0 \\
\vdots & \vdots & \vdots & \ddots & \vdots \\
0 & 0 & 0 & \cdots & 0
\end{bmatrix}
\begin{bmatrix}
E_k \\
E_{k-1} \\
E_{k-2} \\
\vdots \\
E_{k-n}
\end{bmatrix}. \quad (3.20)
$$

For convenience, we denote the above equality as $\mathcal{E}_{k+1} = \mathcal{H} \, \mathcal{E}_k$, where $\mathcal{H}$ is the companion block matrix. Taking the 1-norm of the error vectors, we have

$$\|\mathcal{E}_{k+1}\|_1 = \sum_{i=k-n+1}^{i=k+1} \|E_i\|_1 \quad (3.21)$$

and

$$\|\mathcal{E}_k\|_1 = \sum_{i=k-n}^{i=k} \|E_i\|_1. \tag{3.22}$$

Also, if $\max\left\{\sum_{i=k-n}^{i=k} \|(\mathcal{C}_i)_p\|_1, p = 1, \ldots, N\right\} \geq 1$, then

$$\|\mathcal{H}\|_1 = \max\left\{\sum_{i=1}^{i=k} \|(\mathcal{C}_i)_p\|_1, p = 1, \ldots, N\right\}, \tag{3.23}$$

else if $\max\left\{\sum_{i=k-n}^{i=k} \|(\mathcal{C}_i)_p\|_1, p = 1, \ldots, N\right\} < 1$, we then have $\|\mathcal{H}\|_1 = 1$. Therefore, if $\max\left\{\sum_{i=k-n}^{i=k} \|(\mathcal{C}_i)_p\|_1, p = 1, \ldots, N\right\} < 1$, then $\|\mathcal{E}_{k+1}\|_1 \leq \|\mathcal{E}_k\|_1$. ∎

Remark 3.8. Two different monotonic convergence conditions have been defined depending on Γ_{k+1}. Theorem 3.3 covers the case when $\Gamma_{k+1} \neq 0$, and Theorem 3.4 and Theorem 3.7 cover the cases when $\Gamma_{k+1} = 0$.

Remark 3.9. In Theorem 3.3, the MC condition is defined in the l_2-norm topology, while in Theorem 3.4, the MC is defined in the general l_p-norm topology. In Theorem 3.7, the MC condition is defined in the l_1-norm topology. Specifically, in Theorem 3.7, the BSMC condition $\|\mathcal{E}_{k+1}\|_1 \leq \|\mathcal{E}_k\|_1$ is equivalent to $\|E_{k+1}\|_1 \leq \|E_1\|_1$ (and, by the properties of matrix 1-norms and ∞-norms, the MC condition in Theorem 3.7 can also be extended to the l_∞-norm topology).

In Theorem 3.4 and Theorem 3.7, we defined two different MC conditions for HOILC when $\Gamma_{k+1} = 0$. Theorem 3.7 is much less conservative than Theorem 3.4. However, it is still difficult to guarantee the MC of HOILC systems. The following discussion shows our concern.

Let us assume that $\Gamma_{k+1} = 0$, Λ_i are lower Toeplitz matrices, and $\Gamma_i, i = k-n, \ldots, k-1$ have the same structure as Γ_k. By Theorem 3.7, the BSMC condition for MC of HOILC is given as $\max\left\{\sum_{i=k-n}^{i=k} \|(\mathcal{C}_i)_p\|_1, p = 1, \ldots, N\right\} < 1$. Let us check each row vector of $\mathcal{C}_i$. In general, we consider the hth row vector as

$$\sum_{i=k-n}^{i=k} \|(\mathcal{C}_i)_h\|_1 = \|(\mathcal{C}_k)_h\|_1 + \cdots + \|(\mathcal{C}_{k-n})_h\|_1, \tag{3.24}$$

where the subscript h means the hth row vector and the subscript 1 means the l_1-norm. Let us have the following relationship:

$$\|(\mathcal{C}_k)_h\|_1 + \cdots + \|(\mathcal{C}_{k-n})_h\|_1 \geq \|(\Lambda_k + \cdots + \Lambda_{k-n})_h \\ - (H\Gamma_k + \cdots + H\Gamma_{k-n})_h\|_1. \tag{3.25}$$

From the condition $\Lambda_k + \cdots + \Lambda_{k-n} = I$ and using $H\Gamma_k + \cdots + H\Gamma_{k-n} = H(\Gamma_k + \cdots + \Gamma_{k-n})$, we change $\|(\Lambda_k + \cdots + \Lambda_{k-n})_h - (H\Gamma_k + \cdots + H\Gamma_{k-n})_h\|_1$

as $\|I_h - (H(\Gamma_k + \cdots + \Gamma_{k-n}))_h\|_1$. Here, we replace $\Gamma_k + \cdots + \Gamma_{k-n}$ by Γ, because Γ_i have the same structure, so Γ can represent $\Gamma_k + \cdots + \Gamma_{k-n}$ with the same structure. Then, $\|I_h - (H(\Gamma_k + \cdots + \Gamma_{k-n}))_h\|$ is the l_1-norm of the hth row vector of $I - H\Gamma$. Therefore, since

$$\|I - H\Gamma\|_1 = \max_{h=1,\cdots,N} \left\{ \|I_h - (H(\Gamma_k + \cdots + \Gamma_{k-n}))_h\|_1 \right\}, \qquad (3.26)$$

if $\|I - H\Gamma\|_1 > 1$, then $\sum_{i=k-n}^{i=k} \|(\mathcal{C}_i)_h\|_1$ is always bigger than 1. Thus, if $\|I - H\Gamma\|_1 > 1$, then the MC in the HOILC is not guaranteed in the l_1-norm topology.

Remark 3.10. If $\Gamma_{k+1} = 0$ and MC is not achieved in FOILC, it is difficult to guarantee MC in the HOILC, because, generally, $\Gamma_i, i < k$, have the same structure as Γ_k, and Λ_i are designed as Toeplitz matrices. Also note that the maximum singular value of the companion matrix $\mathcal{H}$ is bigger than 1 (i.e., $\bar{\sigma}(\mathcal{H}) \geq 1$), because $\bar{\sigma}(A)$ is always bigger than $\max\{|a_{ij}|\}$.

3.3 Chapter Summary

In this chapter, we provided a brief technical overview of the super-vector approach to ILC. AS and MC conditions were derived for HOILC systems. We noted that in general it is very difficult to find an analytical, monotonically-convergent condition for HOILC systems. For this reason, in the following chapters of this monograph, we mainly focus on FOILC systems if there is no special indication otherwise.

Part II

Robust Interval Iterative Learning Control

4

Robust Interval Iterative Learning Control: Analysis

This chapter considers the stability analysis of ILC systems when the plant Markov parameters are subject to interval uncertainties. Using the super-vector approach to ILC, vertex Markov matrices are employed to develop asymptotic stability (AS) and monotonic convergence (MC) conditions for the ILC process. It is shown that Kharitonov segments between vertex matrices are not required for checking the stability of interval SVILC systems, but instead, checking just the vertex Markov matrices is enough.

4.1 Interval Iterative Learning Control: Definitions

The advantage of the super-vector notation is that the 2-dimensional problem of ILC is changed to a 1-dimensional multi-input, multi-output (MIMO) problem [298, 308, 300]. Using the 1-dimensional input–output relationship $Y_k = HU_k$ and considering the discrete-time ILC update law $U_{k+1} = U_k + \Gamma E_k$, the resulting "closed-loop system" in the iteration domain is given by $E_{k+1} = (I - H\Gamma)E_k$ where $E_k = Y_d - Y_k$. Thus, the problem becomes the design of the matrix Γ, given the system Markov matrix H, so that the error signal E_k converges (AS) or converges monotonically (MC).

In the ILC literature, analogous to the classical control literature, robust design of the learning gain matrix has been considered using standard techniques such as H_∞-ILC [360, 102], LQ-ILC [136], optimal-ILC [9, 348], etc., along the time-domain axis. However, though the classical control literature has also considered the problem of robust control for plants subject to parametric interval uncertainty, to date there has been little research on the topic of interval uncertainty in the ILC literature (that is, ILC when the system matrix H is an interval matrix). In this section, the stability problem for first-order ILC (FOILC) systems will be studied when the plant Markov parameters are subject to parametric interval uncertainties.

In the robust control literature, there are numerous results related to Hurwitz stability for interval matrices [220, 362] and Schur stability [39, 388].

Kharitonov's theorem has also been very popular for interval matrix stability analysis, e.g., [42, 243]. However, all these works require a significant amount of calculation and cannot be directly applied for checking the MC of an interval ILC (IILC) system. Thus, in this section, we give analytical solutions for checking the convergence properties of the IILC system. Similar to the Kharitonov vertex polynomial method, it will be shown that the extreme values of the interval Markov parameters provide a monotonic convergence condition for the interval ILC system.

Let the ILC learning gain matrix Γ be given as

$$\Gamma = [\gamma_{ij}], \ i, j = 1, \ldots, n, \tag{4.1}$$

where the gains γ_{ij} are the elements of Γ. The gains are called Arimoto-like if $\gamma_{ij} = 0$ for $i \neq j$ and $\gamma_{ij} = \gamma$ for $i = j$. The gains are called causal ILC gains for $i > j$, and the gains are called noncausal ILC gains for $i < j$. If the gain matrix does not exhibit Toeplitz-like symmetry it is called a time-varying learning algorithm. The learning gain matrix Γ could be fully populated or partially populated. If a limited number of causal and noncausal "bands"[1] are used, it is called a "band-limited" gain matrix. In ILC, as described in Chapter 2, there are two stability concepts defined on the iteration domain: AS and MC. Note, however, these AS and MC concepts are defined iteration-wise and should be carefully distinguished from the AS and MC of traditional control schemes that are defined trajectory-wise in the time domain. To emphasize the difference from the traditional definitions, the following formal definitions are provided:

Definition 4.1. *Let us assume that the desired output trajectory $y_d(t)$, $t = 1, \ldots, N$, is given and the controlled system output trajectory $y_k(t)$, $t = 1, \ldots, N$ is measured, where k is iteration trial number. From these signals we form the super-vectors Y_d and Y_k, respectively. Then, the ILC system is:*

- *Stable if, for each $\epsilon > 0$, there is a $\delta = \delta(\epsilon) > 0$ such that*

$$\|Y_d - Y_1\| < \delta \ \Rightarrow \|Y_d - Y_k\| < \epsilon, \ \forall \ k > \overline{k} \ \text{ for some } \overline{k}.$$

- *Asymptotically stable (AS) if it is stable and $\|Y_d - Y_k\| \to 0$ as $k \to \infty$, for all $\|Y_d - Y_1\| < c$, where c is a positive constant.*
- *Monotonically-convergent (MC) if it is stable and $\|Y_d - Y_{k+1}\| < \|Y_d - Y_k\|$, for all k.*
- *Semi monotonically convergent if it is stable and $\|Y_d - Y_{k+1}\| \leq \|Y_d - Y_k\|$, for all k.*
- *Monotonically asymptotically stable if it is asymptotically stable and $\|Y_d - Y_{k+1}\| < \|Y_d - Y_k\|$, for all k.*

∎

---

[1] A "band" means a triangular segment of the matrix that includes the main diagonal.

In the definitions given above, if only stability or monotonic convergence is guaranteed, the final error of the ILC process will not be guaranteed to be zero, which leads us to define the notion of the baseline error of the ILC process as follows:

Definition 4.2. *For ILC systems that are stable or MC, but not AS, let $q = q(c) > 0$, where $c > 0$ is the smallest possible constant such that*

$$\|Y_d - Y_k\| < c, \quad \forall\, k > q.$$

Then the constant c is called the baseline error of the ILC process. ∎

Throughout the monograph, we focus on the design of monotonically-convergent (MC) ILC systems. However, note that in interval ILC (IILC) systems, which will be studied in this chapter, Chapter 5, and Chapter 6, we also guarantee zero steady-state error as $k \to \infty$ except the case of using the so-called Q-filter as described in Section 6.4. Therefore, in IILC, we guarantee monotonically asymptotically stable ILC systems in most cases. In Chapter 8, we study MC conditions for stochastic ILC systems in a stochastic sense (see Section 8.1.2). However, we also note that in Chapter 8, MC conditions will be equivalent to monotonically asymptotically stable ILC conditions if stochastic disturbances and noise are not counted. Thus, in this monograph, even though we simply say MC ILC, the system will be in fact monotonically asymptotically stable ILC. Thus, unless otherwise noted, throughout the monograph, MC means monotonically asymptotically stable.

Now, based on Section 3.1 and Section 3.2, for the FOILC system the following equivalent conditions for AS and MC can be given:

Definition 4.3. *Assuming a fully populated learning gain matrix:*

- *A necessary and sufficient condition for AS is $\rho(I - H\Gamma) < 1$, where $\rho(\cdot)$ is the spectral radius of the matrix $(\cdot)$.*
- *A sufficient condition for MC is $\|I - H\Gamma\| < 1$. In particular, the MC condition in the 1-norm topology is $\|I - H\Gamma\|_1 < 1$, in the 2-norm topology is $\|I - H\Gamma\|_2 < 1$, and in the ∞-norm topology is $\|I - H\Gamma\|_\infty < 1$.*

∎

Remark 4.4. Clearly, $\|I - H\Gamma\| < 1$ is only a sufficient condition, because as a counterexample, if E_k is in the null space of $I - H\Gamma$, then $E_{k+i} = 0$ for all $i \geq 1$. Also, it is always possible to make $I - H\Gamma$ such that $I - H\Gamma$ has a null space, because $I - H\Gamma$ can be a singular matrix if we select γ_1 to be $\gamma_1 = 1/h_1$.

We comment that if time-invariant Arimoto-like gains $\gamma_i = \gamma$ are used for designing Γ, a simplified AS condition can be given as $|1 - \gamma h_1| < 1$ where h_1 is the first nonzero Markov parameter. This is the main reason why ILC systems do not require information about the system matrix A, because $h_1 = CB$ does not depend on A.

Next, to clarify the concepts we will use for the robust stability analysis, the following definitions are introduced:

Definition 4.5. *Interval scalars, interval matrices, upper-bound, lower-bound, and vertex matrices, and interval Markov matrices are defined as follows:*

- *A scalar a is called an interval parameter if it lies between two extreme boundaries according to $a \in [\underline{a}, \overline{a}]$, where $\underline{a}$ is the minimum value of a and $\overline{a}$ is the maximum value of a. To represent the fuzzy characteristics of an interval parameter, the superscript I is used to denote an interval scalar as follows: $a \in a^I := [\underline{a}, \overline{a}]$.*
- *An interval matrix (A^I) is defined as a set of matrices whose entries are interval scalars:*

$$A^I = \left\{ A : \; A = \left[a_{ij} \in [\underline{a_{ij}}, \overline{a_{ij}}] \right], i, j = 1, \cdots, n \right\}, \qquad (4.2)$$

  where $\overline{a_{ij}}$ is the maximum extreme value of the ith row and j^{th} column element of the interval matrix and $\underline{a_{ij}}$ is the minimum extreme value of the ith row and j^{th} column element of the interval plant.
- *The upper bound matrix $(\overline{A})$ is a matrix whose elements are $\overline{a_{ij}}$. The lower bound matrix $(\underline{A})$ is a matrix whose elements are $\underline{a_{ij}}$. The vertex matrices (A^v) are defined by*

$$A^v = \left\{ A : \; A = \left[a_{ij} \in \{\underline{a_{ij}}, \overline{a_{ij}}\} \right], i, j = 1, \cdots, n \right\}. \qquad (4.3)$$

- *If the Markov parameters of a system are interval scalars $h_i \in [\underline{h_i}, \overline{h_i}]$, then we say the system is an interval plant and its interval Markov matrix is denoted as H^I, with associated upper-bound, lower-bound, and vertex Markov matrices $\overline{H}$, $\underline{H}$ and H^v, respectively, defined from $h_i \in \{\underline{h_i}, \overline{h_i}\}$.* ∎

To continue we must also define arithmetic for interval parameters. The following are standard definitions from the literature (refer to [314, 8, 215]). For nonempty closed intervals,

- Addition of two real interval scalars x^I and y^I is defined and calculated as $x^I \oplus y^I = [\underline{x} + \underline{y}, \overline{x} + \overline{y}]$.
- Substraction is defined and calculated as $x^I \ominus y^I = [\underline{x} - \overline{y}, \overline{x} - \underline{y}]$.
- Multiplication is defined and calculated as

$$x^I \otimes y^I = \left[\min\{\underline{x}\underline{y}, \underline{x}\overline{y}, \overline{x}\underline{y}, \overline{x}\overline{y}\}, \max\{\underline{x}\underline{y}, \underline{x}\overline{y}, \overline{x}\underline{y}, \overline{x}\overline{y}\} \right]. \qquad (4.4)$$

- Division should be carefully defined, based on [215], by first defining the inverse operator as

$$1/x^I = \emptyset \text{ iff } x^I = [0,0]$$
$$= [1/\overline{x}, 1/\underline{x}] \text{ iff } 0 \notin x^I$$
$$= [1/\overline{x}, \infty) \text{ iff } \underline{x} = 0 \text{ and } \overline{x} > 0$$
$$= (-\infty, 1/\overline{x}] \text{ iff } \underline{x} < 0 \text{ and } \overline{x} = 0$$
$$= (-\infty, \infty) \text{ iff } \underline{x} < 0 \text{ and } \overline{x} > 0.$$

Based on the definition of the inverse operator, division of two interval scalars is then simply defined and calculated as $x^I \oslash y^I = x^I \otimes \frac{1}{y^I}$.

Using the interval arithmetic given above for scalar intervals, multiplication and summation between interval matrices can be defined such as $A^I = B^I \oplus C^I$ and $A^I = B^I \otimes C^I$ in the usual way.

4.2 Robust Stability of Interval Iterative Learning Control

The interval ILC (IILC) problem is concerned with the analysis and design of the ILC system when the system to be controlled is subject to structured uncertainties in its Markov parameters. There are two classes of problems. First, given an interval Markov matrix H^I and a gain matrix Γ, what are the stability and convergence properties of the closed-loop system? Second, given an interval Markov matrix H^I, design Γ so as to achieve desired stability and convergence properties of the closed-loop system. The second problem will be discussed in Chapters 5 and 6. In the following sections of this chapter, the first problem is considered. We first consider conditions for AS and MC. We then discuss a singular value approach to the stability analysis problem. Next the robust stability of higher-order ILC (HOILC) is addressed. Finally, we conclude with experimental results illustrating the theoretical concepts developed in the chapter.

4.2.1 Asymptotical Stability of the Interval FOILC

To develop our test for AS of the IILC, the following two lemmas are adopted from the literature.

Lemma 4.6. *For a given interval matrix plant A^I, the spectral radius of $A \in A^I$ is bounded by the maximum value of the spectral radii of the vertex matrices $A \in A^v$.* ∎

Proof. See [202, 177]. ∎

Let us make the following substitutions:

$$s_{ij}^1 := \overline{a_{ij}} \text{ if } i = j; \quad s_{ij}^1 := \max\{|\underline{a_{ij}}|, |\overline{a_{ij}}|\} \text{ if } i \neq j;$$

$$s_{ij}^2 := \underline{a_{ij}} \text{ if } i = j; \quad s_{ij}^2 := \min\{-|\underline{a_{ij}}|, -|\overline{a_{ij}}|\} \text{ if } i \neq j,$$

where s_{ij}^1 is the ith row and j^{th} column element of the matrix S^1; s_{ij}^2 is an element of the matrix S^2, and $a_{ij}^I = [\underline{a_{ij}}, \overline{a_{ij}}]$ is an element of the general interval matrix A^I.

Lemma 4.7. *Let an interval matrix be given as $\underline{A} \leq A \leq \overline{A}$, $A \in A^I$. If $\beta = \max\{\rho(S^1), \rho(S^2)\} < 1$, where ρ is the spectral radius, then the interval matrix A^I is Schur stable.* ∎

Proof. See Theorem 3.2 of [108]. ∎

Let us consider the case of a general learning gain matrix Γ that consists of causal and noncausal gains and let us suppose that in $I - H\Gamma$ the Markov matrix is assumed to be uncertain according to $H \in H^I$. Thus, the lower boundary and the upper boundary of $I - H\Gamma$, $H \in H^I$ should be recalculated. For convenience, letting $T = H\Gamma$, and using interval arithmetic, we have $T^I = H^I \otimes \Gamma$. Further defining $P := I - T$ and $P^I := I - T^I$, the lower and upper boundaries of P^I, i.e., $\underline{P}$ and $\overline{P}$, can be calculated easily using the interval computation software Intlab® [179]. Then, using the lower and upper boundaries of P^I, it can be shown from Lemma 4.6 that the maximum spectral radius of the interval matrix $P^I = I - H^I \otimes \Gamma$ occurs at one of the vertex matrices, P^v, of P^I. However, it is quite messy to check all the vertex matrices. To avoid a large number of computations, Lemma 4.7 can be used, because it requires checking only two matrices, but with a conservative result. Hence, there is a trade-off between using Lemma 4.6 and using Lemma 4.7. Thus, we are led to develop the following theorem, which provides a necessary and sufficient condition for robust AS, with a greatly reduced computational load.

Theorem 4.8. *Let the first Markov parameter h_1 be an interval parameter given by $h_1 \in h_1^I = [\underline{h_1}, \overline{h_1}]$ and let Arimoto-like/causal ILC gains be used in Γ. Then the first-order IILC system, $E_{k+1} = (I - H\Gamma)E_k$, $H \in H^I$, is AS if and only if*

$$\max\left\{|1 - \gamma_{ii}\underline{h_1}|, |1 - \gamma_{ii}\overline{h_1}|\right\} < 1, i = 1, \dots, n. \tag{4.5}$$

∎

Proof. Using the fact that H^I is a lower-triangular Toeplitz matrix and Γ is a lower-triangular matrix, we know that $I - H\Gamma$ is a lower-triangular matrix for all $H \in H^I$. Thus, the diagonal terms of $I - H\Gamma$, given as $\{1 - \gamma_{ii}h_1\}$, $h_1 \in h_1^I$, $i = 1, \cdots, n$, are the eigenvalues of $I - H\Gamma$, $H \in H^I$. When $i = k$, the maximum value of $|1 - \gamma_{kk}h_1|$, $h_1 \in h_1^I$ occurs at one of the $h_1 \in h_1^v = \{\underline{h_1}, \overline{h_1}\}$, because $|1 - \gamma_{kk}h_1|$ is the absolute value of $1 - \gamma_{kk}h_1$, $h_1 \in h_1^I$. Therefore, the maximum of $\{|1 - \gamma_{ii}\underline{h_1}|, |1 - \gamma_{ii}\overline{h_1}|\}$ occurs at one of the $h_1^v = \{\underline{h_1}, \overline{h_1}\}$. Thus, if and only if $\max\{|1 - \gamma_{ii}\underline{h_1}|, |1 - \gamma_{ii}\overline{h_1}|\} < 1$ is satisfied, the system is AS from Definition 4.3. ∎

4.2.2 Monotonic Convergence

Now we consider the MC of the IILC system. To prove our main theorems, the following lemmas are developed first.

Lemma 4.9. *Let $x \in x^I = [\underline{x}, \overline{x}]$ be an interval parameter. Then for*

$$y = |\gamma_{11}x + \gamma_{12}| + |\gamma_{21}x + \gamma_{22}|, \ \forall \gamma_{11}, \gamma_{12}, \gamma_{21}, \gamma_{22} \in \mathcal{R}, \qquad (4.6)$$

the $\max\{y\}$ *occurs, for all* $x \in x^I$, *at a vertex point of* x^I *(i.e.,* $x \in x^v = \{\underline{x}, \overline{x}\}$*).* ∎

Proof. Figure 4.1 shows line drawings of $|\gamma_{11}x + \gamma_{12}|$ and $|\gamma_{21}x + \gamma_{22}|$. Let us check the three different regions: $R_1 \in [\underline{x}, \underline{x_0}]$, $R_2 \in [\underline{x_0}, \overline{x_0}]$, and $R_3 \in [\overline{x_0}, \overline{x}]$. In region R_1, $\max\{y\}$ occurs at $\underline{x}$, because $y_1 + y_2 > y_3 + y_4$. Also, in region R_3, $\max\{y\}$ occurs at $\overline{x}$, because $y_8 + y_9 > y_6 + y_7$. Now consider R_2. In region R_2, y is the just summation of two linear straight lines (i.e, the line connecting from y_6 to y_4 and the line connecting from y_3 to y_7) like $y = \gamma_{11}x + \gamma_{12} + \gamma_{21}x + \gamma_{22} = (\gamma_{11} + \gamma_{21})x + \gamma_{12} + \gamma_{22}$, which is represented by line l_1. Thus, in region R_2, the value of y linearly increases or linearly decreases. Hence, $\max\{y\}$ in R_2 occurs at $x \in \{\underline{x_0}, \overline{x_0}\}$. Finally, from Figure 4.1, since it is true that $\max\{y_1 + y_2, y_8 + y_9\} > \max\{y_3 + y_4, y_6 + y_7\}$, the proof is completed. ∎

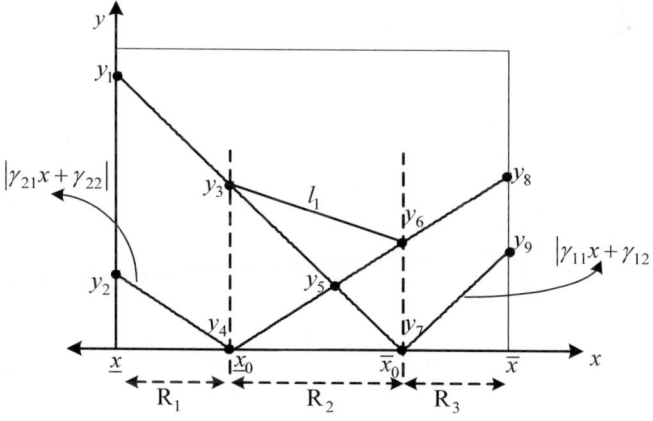

Fig. 4.1. Supplementary drawing for the proof of Lemma 4.9

Lemma 4.10. *Let $x \in x^I = [\underline{x}, \overline{x}]$ be an interval parameter. Then for*

$$y = |\gamma_{11}x + \gamma_{12}| + |\gamma_{21}x + \gamma_{22}| + $$
$$\cdots + |\gamma_{n1}x + \gamma_{n2}|, \ \forall \gamma_{i1}, \gamma_{i2} \in \mathcal{R}, i = 1, \ldots, n, \qquad (4.7)$$

the $\max\{y\}$ *occurs, for all* $x \in x^I$, *at one of vertex points of* x^I. ∎

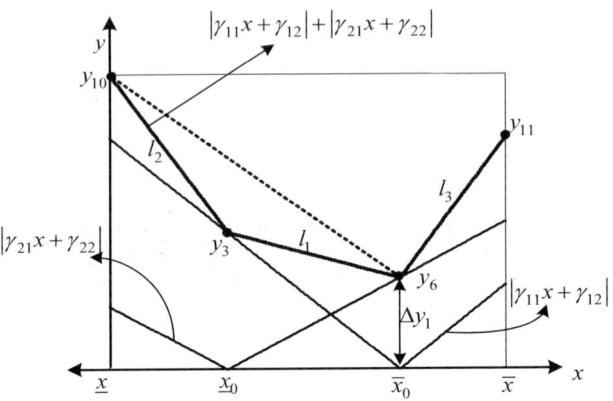

Fig. 4.2. Supplementary drawing for the proof of Lemma 4.10

Proof. From Figure 4.1, consider R_2 again. In region R_2, the value of y (summation of two lines) is l_1. Now consider Figure 4.2, where $y = |\gamma_{11}x + \gamma_{12}| + |\gamma_{21}x + \gamma_{22}|$ is represented by lines l_1, l_2, and l_3. To prove Lemma 4.10, draw a supplementary line (the dashed line in the figure) from point y_{10} to point y_6. Then, the line connecting y_{10} and y_6 and the line connecting y_6 and y_{11} can be represented by a line that we can write as $y = |\gamma_1^1 x + \gamma_2^1| + \triangle y_1$, with $\gamma_1^1, \gamma_2^1 \in \mathcal{R}$, and $\triangle y_1 \in \mathcal{R}^+$. Note that this approach does not change the result, because the triangular area included by points y_3, y_6, and y_{10} does not add any value to vertex point values (i.e., y_{10} and y_{11}) whereas if the maximum value still occurs at a vertex point after the triangular area is added, it is certain that the maximum value always occurs at a vertex point. Now, let us check the following:

$$y = |\gamma_{11}x + \gamma_{12}| + |\gamma_{21}x + \gamma_{22}| + |\gamma_{31}x + \gamma_{32}|, \qquad (4.8)$$

which is rewritten as $y = |\gamma_1^1 x + \gamma_2^1| + \triangle y_1 + |\gamma_{31}x + \gamma_{32}|$. Here, ignore $\triangle y_1$, because $\triangle y_1$ is a constant value at all x (i.e., for all $x \in [\underline{x}, \overline{x}]$). Then, from Lemma 4.9, the maximum value of y occurs at a vertex of x (i.e., $x \in \{\underline{x}, \overline{x}\}$).

For the general case, we proceed by induction. Let us assume that, for $n = m$, there exist γ_1^{m-1}, γ_2^{m-1}, and $\triangle y_{m-1}$ such that $y = |\gamma_{11}x + \gamma_{12}| + |\gamma_{21}x + \gamma_{22}| + \cdots + |\gamma_{m1}x + \gamma_{m2}| = |\gamma_1^{m-1}x + \gamma_2^{m-1}| + \triangle y_{m-1} + |\gamma_{m1}x + \gamma_{m2}|$. Let us check the case $n = m + 1$. We have

$$y = |\gamma_{11}x + \gamma_{12}| + |\gamma_{21}x + \gamma_{22}| + $$
$$\cdots + |\gamma_{m1}x + \gamma_{m2}| + |\gamma_{(m+1)1}x + \gamma_{(m+1)2}|$$
$$= |\gamma_1^{m-1}x + \gamma_2^{m-1}| + \triangle y_{m-1} + |\gamma_{m1}x + \gamma_{m2}| + |\gamma_{(m+1)1}x + \gamma_{(m+1)2}|.$$
$$\qquad (4.9)$$

Then, ignoring $\triangle y_{m-1}$ and using the same procedure as followed for (4.8), for $n = m + 1$, we know that the maximum value of y occurs at a vertex of x. This completes the proof.   ∎

The following lemma considers multiple interval parameters.

Lemma 4.11. *Let $x^j \in \left[\underline{x^j}, \overline{x^j}\right]$, $j = 1, \ldots, m$, be interval parameters (for convenience we omit the superscript I and v). Then for*

$$y = \left|(\gamma_{11}^1 x^1 + \gamma_{12}^1) + \cdots + (\gamma_{11}^m x^m + \gamma_{12}^m)\right| +$$
$$\cdots + \left|(\gamma_{n1}^1 x^1 + \gamma_{n2}^1) + \cdots + (\gamma_{n1}^m x^m + \gamma_{n2}^m)\right|,$$
$$\forall \gamma_{i1}^j, \gamma_{i2}^j \in \mathcal{R}, i = 1, \ldots, n, j = 1, \ldots, m, \qquad (4.10)$$

the $\max\{y\}$ occurs, for all $x^j \in \left[\underline{x^j}, \overline{x^j}\right]$, at the vertices of the m scalars x^j. ∎

Proof. Lemma 4.10 shows that, in the following equation:

$$y = |\gamma_{11} x + \gamma_{12}| + \cdots + |\gamma_{n1} x + \gamma_{n2}|, \qquad (4.11)$$

$\gamma_{i2}, i = 1, \ldots, n$, can be any real values. Thus, in (4.10), if the following substitution is used:

$$\gamma_{i2}^1 + \cdots + (\gamma_{i1}^m x^m + \gamma_{i2}^m) := \gamma_{i2}, i = 1, \ldots, n,$$

then (4.10) has the same form as (4.7). Therefore, $\max\{y\}$ in (4.10) occurs at a vertex point of x^1 by Lemma 4.10, because all elements of $\{x^j\}$, $j = 1, \ldots, m$, are independent of one another. Next, let us place $\gamma_{i1}^j x^j + \gamma_{i2}^j, j \in \{1, \ldots, m\}$, $i = 1, \ldots, n$ in the front of each absolute value term according to

$$y = \left|(\gamma_{11}^j x^j + \gamma_{12}^j) + \sum_{k=1, k\neq j}^{m} (\gamma_{11}^k x^k + \gamma_{12}^k)\right| +$$
$$\cdots + \left|(\gamma_{n1}^j x^j + \gamma_{n2}^j) + \sum_{k=1, k\neq j}^{m} (\gamma_{n1}^k x^k + \gamma_{n2}^k)\right|. \qquad (4.12)$$

By writing $\gamma_{12}^j + \sum_{k=1, k\neq j}^{m} (\gamma_{l1}^k x^k + \gamma_{l2}^k) := \xi_l$, where $l = 1, \ldots, n$, the right-hand side of above equation is changed to

$$y = \left|\gamma_{11}^j x^j + \xi_1\right| + \cdots + \left|\gamma_{n1}^j x^j + \xi_n\right|, \qquad (4.13)$$

where $\xi_i, i = 1, \ldots, n$, could be any real values. This is the same form as (4.7), so the maximum value of y of (4.13) occurs at a vertex point of x^j (i.e.,$x^j \in \{\underline{x^j}, \overline{x^j}\}$) by Lemma 4.10. Here, note that the maximum value of y, which occurs at a vertex point of x^j, is just with respect to x^j. Let us denote this maximum value as y_j^*. Now, it is required to show that the maximum value of y with respect to all intervals (i.e.,$\{x^j\}, j = 1, \ldots, m$) occurs at one of the vertex vectors given by

$$X^v = \left[\{ \underline{x^1}, \overline{x^1} \}, \{ \underline{x^2}, \overline{x^2} \}, \ldots, \{ \underline{x^m}, \overline{x^m} \} \right]. \qquad (4.14)$$

Denote this maximum value as $\overline{y}^*$. Note that $\overline{y}^* \neq y_j^*$. Thus, it is necessary to prove that, when the maximum value of y occurs at a vertex of x^j with fixed j, other interval parameters (i.e., $x^k, k \neq j$,) should be at vertices as well (in this case, $\overline{y}^* = y_j^*$). Even though the maximum value of y occurs at a vertex of x^j, the other intervals $x^k, k \neq j$, might not be at vertex points (in this case, $\overline{y}^* \neq y_j^*$). Next, let us assume that, when the maximum value of y occurs at a vertex of x^j, the other interval parameter $x^k, k \neq j$, is not at a vertex point (i.e., x^k is an element of open set $x^k \in (\underline{x^k}, \overline{x^k})$). Let us change (4.13) using $\xi_i := \gamma_{i1}^k x^k + \xi_i'$ to be

$$y = \left| \gamma_{11}^k x^k + \gamma_{11}^j x^j + \xi_1' \right| + \left| \gamma_{21}^k x^k + \gamma_{21}^j x^j + \xi_2' \right| +$$
$$\cdots + \left| \gamma_{n1}^k x^k + \gamma_{n1}^j x^j + \xi_n' \right|, \qquad (4.15)$$

where $\xi_i', i = 1, \ldots, n$, could be any real values. Because (4.15) and (4.13) are the same equations, the maximum value of y still occurs at a vertex of x^j. Thus, $\max\{y\} = y_j^*$, but $\max\{y\} \neq y_k^*$ and $\max\{y\} \neq \overline{y}^*$, where y_k^* is the maximum value with respect to x^k. However, by Lemma 4.10, y of (4.15) can be maximized more with respect to x^k. In other words, even though the current maximum value of (4.15) is y_j^*, when x^k is at one of vertex points, y_j^* can be increased more. Just by comparing the following two values:

$$y = \begin{cases} y_j^* , & \text{if } x^k \in \left(\underline{x^k}, \overline{x^k} \right), \ x^j = \left\{ \underline{x^j}, \overline{x^j} \right\} \\ y_{jk}^* , & \text{if } x^k \in \left\{ \underline{x^k}, \overline{x^k} \right\}, \ x^j = \left\{ \underline{x^j}, \overline{x^j} \right\} \end{cases} \qquad (4.16)$$

it is found that $\max\{y_{jk}^*\} \geq \max\{y_j^*\}$ by Lemma 4.10. Then, the maximum value of y of (4.15) with respect to k and j occurs at one of $\left\{ \left\{ \underline{x^k}, \overline{x^k} \right\}, \left\{ \underline{x^j}, \overline{x^j} \right\} \right\}$. Finally, since $k \in \{1, \cdots, m\}$, the following is true by induction:

$$\max\{y\} = y_{123\cdots m}^*, \quad \text{when } x^i = \left\{ \underline{x^i}, \overline{x^i} \right\}, i = 1, \ldots, m, \qquad (4.17)$$

where $y_{123\cdots m}^*$ is the maximum value with respect to all interval parameters. Then, from the relationship $\overline{y}^* = y_{123\cdots m}^*$, the maximum value of y occurs at one of the vertex vectors:

$$X^v = \left[\{ \underline{x^1}, \overline{x^1} \}, \{ \underline{x^2}, \overline{x^2} \}, \ldots, \{ \underline{x^m}, \overline{x^m} \} \right].$$

Thus, the proof is completed.   ∎

Next, using these lemmas, the following theorems can be proven.

Theorem 4.12. *Given interval Markov parameters* $h_i \in h_i^I = [\underline{h_i}, \overline{h_i}]$*, the IILC system is MC in the l_∞-norm topology if*

$$\max\{\|I - H\Gamma\|_\infty, \ \forall H \in H^v\} < 1, \tag{4.18}$$

where H^v are vertex Markov matrices of the interval plant. ∎

Proof. Based on Definition 4.3, the theorem can be proved by showing that $\max\{\|I - H\Gamma\|_\infty, \forall H \in H^I\} = \max\{\|I - H\Gamma\|_\infty, \forall H \in H^v\}$. Let us expand $I - H\Gamma$ as

$$I_{n\times n} - \begin{bmatrix} h_1 & 0 & \cdots & 0 \\ h_2 & h_1 & \cdots & 0 \\ \vdots & \vdots & \ddots & \vdots \\ h_n & h_{n-1} & \cdots & h_1 \end{bmatrix} \begin{bmatrix} \gamma_{11} & \gamma_{12} & \cdots & \gamma_{1n} \\ \gamma_{21} & \gamma_{22} & \cdots & \gamma_{2n} \\ \vdots & \vdots & \ddots & \vdots \\ \gamma_{n1} & \gamma_{n2} & \cdots & \gamma_{nn} \end{bmatrix}.$$

The row vectors of $I - H\Gamma$ are expressed as

$$(I - H\Gamma)_1 = [1 - h_1\gamma_{11}, -h_1\gamma_{12}, \ldots, -h_1\gamma_{1n}]$$
$$(I - H\Gamma)_2 = [-(h_2\gamma_{11} + h_1\gamma_{21}), 1 - (h_2\gamma_{12} + h_1\gamma_{22}), \ldots, -(h_2\gamma_{1n} + h_1\gamma_{2n})]$$

$$\vdots$$

$$(I - H\Gamma)_n = [-(h_n\gamma_{11} + \cdots + h_1\gamma_{n1}), -(h_n\gamma_{12} + \cdots + h_1\gamma_{n2}),$$
$$\ldots, 1 - (h_n\gamma_{1n} + \cdots + h_1\gamma_{nn})], \tag{4.19}$$

where $(I - H\Gamma)_i$ is the ith row vector. Then, we know that $\|I - H\Gamma\|_\infty$ is a function of h_i and γ_{ij}, because the following is true:

$$\|I - H\Gamma\|_\infty = \max\{\|(I - H\Gamma)_1\|_1, \|(I - H\Gamma)_2\|_1, \ldots, \|(I - H\Gamma)_n\|_1\}, \tag{4.20}$$

where $\|(I - H\Gamma)_i\|_1$ is the l_1-norm of each row vector. Assuming fixed ILC gains γ_{ij}, $\|I - H\Gamma\|_\infty$ is expressed in the general form

$$\|I - H\Gamma\|_\infty = |-(h_i\gamma_{11} + \cdots + h_1\gamma_{i1})| +$$
$$\cdots + |1 - (h_i\gamma_{1i} + \cdots + h_1\gamma_{ii})| +$$
$$\cdots + |-(h_i\gamma_{1n} + \cdots + h_1\gamma_{in})|, \tag{4.21}$$

where i means the ith row. Thus, we know that (4.21) is the same form as (4.10) of Lemma 4.11. Note, in Lemma 4.11, x^j are intervals with $\forall \gamma_{i1}^j, \gamma_{i2}^j \in \mathcal{R}, i = 1, \ldots, n, j = 1, \ldots, m$, and in (4.21), h_i are intervals with $\forall \gamma_{ij} \in \mathcal{R}, i, j = 1, \ldots, n$. Therefore, from Lemma 4.11, we conclude that the maximum of $\|I - H\Gamma\|_\infty, \forall H \in H^I$, occurs at one of vertex Markov matrices of the plant. Finally, from this result, the following relationship is true:

$$\max\{\|I - H^I\Gamma\|_\infty, \forall H \in H^I\} = \max\{\|I - H^v\Gamma\|_\infty, \forall H \in H^v\}.$$

This completes the proof. ∎

Theorem 4.13. *Given interval Markov parameters $h_i^I \in [\underline{h_i}, \overline{h_i}]$, the following equality is true:*

$$\max\{\|I - H\Gamma\|_1, \forall H \in H^I\} = \max\{\|I - H\Gamma\|_1, \forall H \in H^v\}, \quad (4.22)$$

where $\|\cdot\|_1$ is the matrix 1-norm, defined as $\|A\|_1 = \max_{j=1,\cdots,n} \sum_{i=1}^{n} |a_{ij}|$. ∎

Proof. The proof can be completed using the same procedure as in the proof of Theorem 4.12. ∎

Remark 4.14. The strong point of Theorem 4.12 and Theorem 4.13 is that they reduce the computational effort significantly. Using the notation $P^I = I - T^I$, it is required to check 2^{n^2} vertex matrices of P^I, while Theorem 4.12 and Theorem 4.13 require only 2^n matrices for the MC test of IILC.[2]

4.2.3 Singular Value Approach

In the preceding subsections 1- and ∞-norms were used for IILC robust stability analysis. In the ILC literature, the maximum singular value, which is a 2-norm, has been widely used as a monotonic convergence criteria. We comment that the MC condition in the l_2-norm topology can also be checked by Theorem 4.12 and Theorem 4.13. That is, if $\|I - H\Gamma\|_\infty \|I - H\Gamma\|_1 < 1$, $\forall H \in H^v$, then $\|I - H\Gamma\|_2 < 1$, $\forall H \in H$". This argument is supported from the fact that:

$$\|A\|_2^2 \leq \|A\|_1 \|A\|_\infty,$$

where A is a general matrix. Thus, by Theorem 4.12 and Theorem 4.13, the MC property of the IILC system in the l_2-norm topology can be checked. Of course, this result could be overly conservative because it uses an inequality. However, if it is assumed that the extreme values of $I - H^I \otimes \Gamma$ are calculated by interval operations, and hence if the interval ranges of $P^I = I - H^I \otimes \Gamma$ are known, then the MC condition in the 2-norm topology can be directly checked by using the maximum singular value of an interval matrix. In this monograph we provide an algorithm for this computation in Appendix B. Using Algorithm B.1, it is straightforward to find the maximum singular value (2-norm) of the IILC system. This application of the result from Appendix B is omitted due to its simplicity.

4.2.4 Robust Stability of Higher-order Interval Iterative Learning Control

Higher-order ILC (HOILC) has been of special interest in the ILC research community, due to the possibility of improved convergence speed and robustness, as described in [43, 74, 67, 66]. However, the robust stability property

---

[2] Note that although the test we have presented for $\max\{\|I - H\Gamma\|_p, \forall H \in H^I\}$ where $p = 1, \infty$ is not overly conservative, it is only a sufficient check for MC.

of HOILC has not been fully studied. In this subsection, we consider a robust stability condition for uncertain HOILC systems using the super-vector framework. As described in Chapter 2, the stability problem of the interval HOILC system is equivalent to that of the stability of a polynomial matrix system. To consider the robust interval HOILC problem, note from (3.14) that for the HOILC system the following error propagation rule is given as

$$E_{k+1} = (H\Lambda_k H^{-1} - H\Gamma_k)E_k + \cdots + (H\Lambda_{k-n} H^{-1} - H\Gamma_{k-n})E_{k-n}. \quad (4.23)$$

To force zero steady-state error we use the relationship $\sum_{i=0}^{i=n} \Lambda_{k-i} = I$ and we require each Λ_{k-i} to be a diagonal matrix, which gives

$$E_{k+1} = (\Lambda_k - H\Gamma_k)E_k + \cdots + (\Lambda_{k-n} - H\Gamma_{k-n})E_{k-n}. \quad (4.24)$$

Then, including interval uncertainties in H, using interval computations, and writing $A_i^I := \Lambda_{k-n} - H^I \otimes \Gamma_{k-i}$, the robust stability condition for the HOILC system can be resolved from the robust stability of an interval polynomial matrix. For a detailed development and results for this interval polynomial matrix computation, refer to Appendix C. In particular, Theorem C.9 and Theorem C.16 can be used directly to test the robust stability of the higher-order IILC system. This application is straightforward, and hence is also omitted.

4.3 Experimental Test

To present an experimental verification of the monotonic convergence of interval ILC systems, the Quanser SRV02[3] rotary motion system was used. The SRV02 is composed of a rotary servo system, geared DC motor, and potentiometer for angle measurement. The power amplifier used in the test was the UPM-15-03, whose maximum output voltage is 15 volts with a maximum continuous current rating of 3 amps. For data acquisition and control, the terminal board and MultiQ board[4] are necessary. The terminal board is composed of an A/D converter, a D/A converter, and encoder inputs. The data acquisition board measures analog signals from sensors on the plant and converts them to digital signals. Figure 4.3 shows the test setup composed of the Quanser SRV02, power amplifier, terminal board, and Windows®-based WinCon software.

The Quanser SRV02 is a linear system, and for the model identification (to find the impulse response, i.e., the Markov parameters), we used the correlation analysis method, which uses a pseudo-random binary sequence (PRBS) as an input. However, since the Quanser SRV02 is deterministic, we had to include small model disturbances in the WinCon-based MATLAB® program.[5]

---

[3] See http://www.quanser.com/english/html/challenges/fs_chall_rotary_flash.htm.
[4] See http://www.quanser.com/english/html/solutions/fs_soln_hardware.html.
[5] We provided 10 percent uncertainties by changing the differentiator constant, which is connected to the encoder output. See

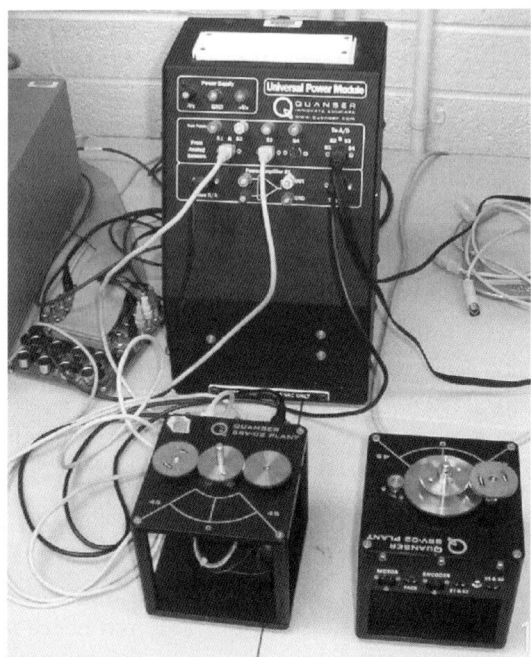

Fig. 4.3. Quanser SRV02 test setups

Thus, the test we describe is a kind of hardware-in-the-loop simulation (HILS). From numerous experimental tests, we empirically found the interval ranges of Markov parameters as shown in Figure 4.4.

Note from Figure 4.4 that after the 31st Markov parameter the interval ranges are ignorable and the magnitude is very small. Thus Markov parameters higher than h_{31} are assumed to be zero. To make $\max\{\|I - H^v\Gamma\|_\infty\} < 1$, we selected the learning gains to be $\gamma_{ij} = 0.4294$ when $i = j$; $\gamma_{ij} = -0.2145$ when $i = j+1$; $\gamma_{ij} = 0.1393$ when $i = j+2$; and $\gamma_{ij} = -0.0883$ when $i = j+3$, which makes $\max\{\|I - H^v\Gamma\|_\infty\} = 0.9522$. In the test, the sampling time was 0.05 seconds and we used 40 discrete-time points. Thus, one iteration trial is executed for two seconds ($0.05 \times 40 = 2$ seconds). Figure 4.5 represents the desired time-periodic repetitive speed trajectory along the time axis. In this figure, the desired speed trajectory in a trial is only for two seconds. Figure 4.5 shows that the system is stopped for a second after two seconds of the repetitive trajectory to make the rotation speed go to zero before starting the next repetition (the zero speed is required for satisfying the initial reset condition). Thus the time-domain plots are shown lasting longer than 40 discrete-time

---

http://www.mathtools.net/MATLAB/Real_Time_and_Embedded/. The WinCon program provides program interfaces between the Simulink®-generated C code with the MULTIQ-3 board. The WinCon client is installed on the host computer with the MULTIQ-3 DACB.

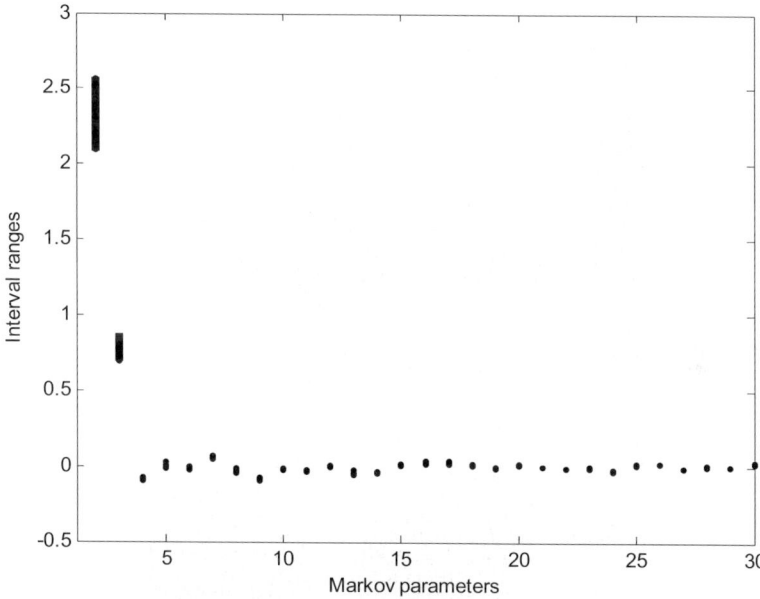

Fig. 4.4. Interval ranges of Markov parameters of Quanser SRV02

points. Figure 4.6 shows one of the interval ILC system processes (that is, for a specific plant in the interval system, we plot the output error for a number of trials). As shown in this plot, the error $(e(t))$ becomes smaller along the time axis, but also as the number of trials increases. This is reinforced by Figure 4.7, which shows the ∞-norms on the iteration domain of several plants from the interval ILC system. The figure shows ten different random plants within the interval uncertainty. The system is monotonically-convergent when the monotonic convergence conditions developed in this chapter are satisfied. However, there are some fluctuations of the responses due to measurement noises and/or looseness of hardware equipment. Likewise, notice that the steady-state error does not go to zero. This is due to the measurement noises of the encoder output and is also representative of the fact that all realistic ILC systems exhibit a baseline error.

4.4 Chapter Summary

In this chapter stability analysis methods for ILC problems were developed for the case when the plant Markov parameters are subject to interval uncertainties. It was shown that checking just the vertex Markov matrices of an interval plant is enough for determining the monotonic convergence properties of the interval ILC system. This is a powerful result from a computational perspective. The results were verified through an experimental test. Also in this

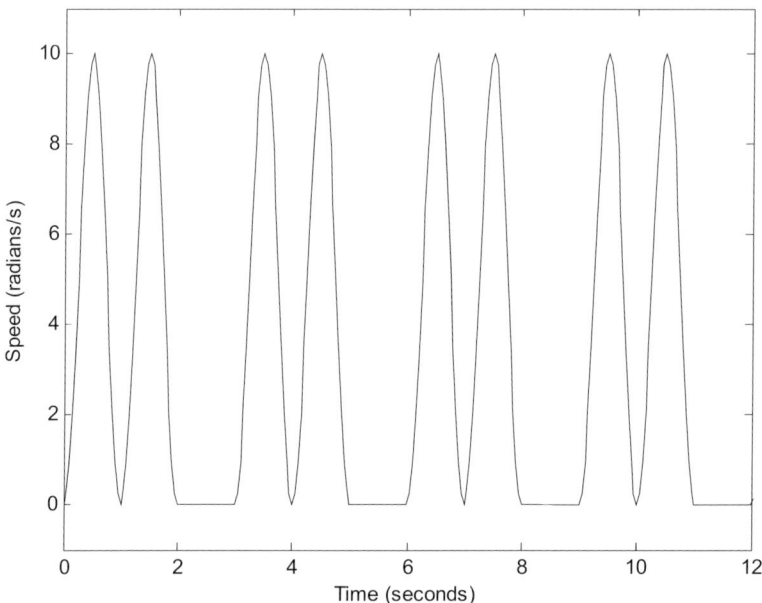

Fig 4.5. Desired time-periodic repetitive speed trajectory

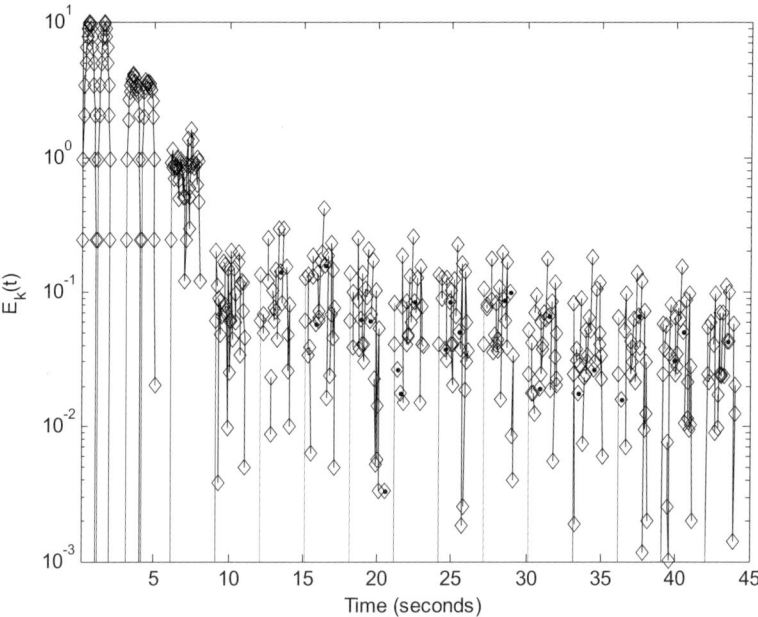

Fig. 4.6. ILC response on the time domain

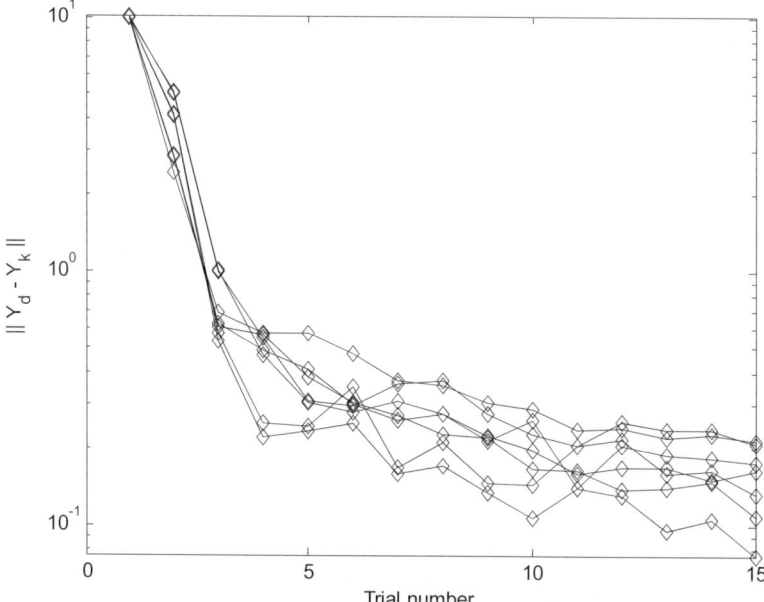

Fig. 4.7. Monotonic convergence of ∞-norm on the iteration domain

chapter, solutions for the maximum singular value of an interval matrix and the robust stability of an interval polynomial matrix system, both detailed in the appendices, were mentioned as methods for testing the robust stability IILC systems. The novelty of the results of this chapter can be summarized as follows:

- First, the existing ILC literature (e.g., [360, 102, 136, 9, 348]) studied MC conditions without considering model uncertainty. However, in this chapter, parametric interval concepts were used for determining a robust MC condition for uncertain ILC systems.
- Second, it was shown that the MC condition of IILC can be checked from vertex Markov matrices, which enables us to establish the MC property of uncertain ILC systems without conservatism. That is, from Lemma 4.9 to Lemma 4.11, we showed that a multiparameter interval optimization problem can be solved using only the extreme vertex points. Then, using these lemmas, in Theorem 4.12 and Theorem 4.13, we established a novel MC condition using the vertex Markov matrices of an uncertain interval ILC system.

Hence, the main contribution of this chapter is to connect ILC theory to parametric robust control theory so that one of the most important practical issues of ILC (i.e., monotonic convergence) can be successfully handled when controlling an uncertain plant.

Finally, we emphasize that in this chapter we have focused on the *analysis* problem for interval ILC problems. In the next chapter we turn our attention to the *synthesis* problem.

5

Schur Stability Radius of Interval Iterative Learning Control

This chapter discusses the synthesis of robust ILC controllers when the plant is subject to interval uncertainties. We show how to use the concept of the Schur stability radius of the IILC system to design robust iterative learning controllers. For the case of parametric or interval perturbations in the system Markov parameters, a discrete Lyapunov equation is used to compute the Schur stability radius of the interval ILC system. After deriving an analytical expression for the IILC stability radius, optimization schemes are suggested to design the learning gain matrix. The proposed approach allows the design of a causal/noncausal, time-varying learning gain matrix.

Of particular interest in ILC, the super-vector approach to ILC has been used to analyze convergence in the iteration domain [308, 300, 302, 303, 311], with AS and MC conditions investigated in [306, 62, 305] and feedback control and quadratic optimal control methods suggested to design the super-vector learning gain matrix in [60, 63, 61, 64]. Recently, an algebraic analysis of the super-vector ILC framework was given in [351] that suggests the possibility of applying interval robustness concepts to the ILC problem to make the system robust against the parameter uncertainties, assuming there exist interval uncertainties in the system Markov parameters. For such a situation, it is desirable to design the ILC learning gain matrix such that the system is stable for the largest possible range of interval uncertainties on the nominal plant. Motivated by this observation, in this chapter we consider two problems. First, in Section 5.1, given a learning gain matrix Γ, we ask: what is the largest interval uncertainty the system can tolerate before it becomes unstable? Second, in Section 5.2, given a particular nominal plant, we seek to find the optimal Γ for which the largest uncertainty interval is as big as possible.

Before proceeding, using the interval concepts given in Chapter 4, let us repeat the following definition for an interval Markov plant:

$$H^I = \left\{ H : \; H = [h_{ij}], \; h_{ij} \in h_{ij}^I := [\underline{h_{ij}}, \overline{h_{ij}}], i, j = 1, \ldots, n \right\}, \qquad (5.1)$$

where $\overline{h_{ij}}$ is maximum extreme value of the ith row, j^{th} column element of the Markov matrix and $\underline{h_{ij}}$ is minimum extreme value of the ith row, j^{th} column element of the Markov matrix. The *upper bound matrix* $(\overline{H})$ is a matrix whose elements are $\overline{h_{ij}}$; and the *lower bound matrix* $(\underline{H})$ is a matrix whose elements are $\underline{h_{ij}}$. Then, from the interval Markov matrix H^I, the nominal Markov matrix and the interval radius matrix are defined as follows.

Definition 5.1. *The nominal Markov matrix is defined by the upper and lower bound matrices as* $H^o = \frac{\overline{H}+\underline{H}}{2}$. *The interval radius matrix is defined as* $\Delta H^r = \frac{\overline{H}-\underline{H}}{2}$. *The maximum value of* ΔH^r *for which a given ILC law has guaranteed AS for all* $H \in H^I$ *satisfying* $-\Delta H^r < H^I - H^o < \Delta H^r$ *is called the maximum Schur asymptotic stability radius and is denoted* ΔH^r_a. *The maximum value of* ΔH^r *for which a given ILC law has guaranteed MC for all* $H \in H^I$ *satisfying* $-\Delta H^r < H^I - H^o < \Delta H^r$ *is called the maximum Schur monotonic stability radius and is denoted* ΔH^r_m. ∎

Consider the first-order ILC (FOILC) update law, given by $U_{k+1} = U_k + \Gamma E_k$. This law gives the evolution of the error vector in iteration domain as $E_{k+1} = (I - H\Gamma)E_k$, where $E_k = Y_d - Y_k$ and Γ is the learning gain matrix. As done in Chapter 4, for convenience, the symbols T and T^I are used, with $T^o = I - H^o\Gamma$, $T = I - H\Gamma$, $H \in H^I$, and $T^I = I - H^I \otimes \Gamma$. Then, the following notation is defined:

$$\Delta T = I - H^o\Gamma - (I - H\Gamma) = (H - H^o)\Gamma = \Delta H\Gamma, \qquad (5.2)$$

where ΔT is the interval uncertainty of the ILC system and ΔH is the interval uncertainty of the nominal Markov matrix. Using ΔT, optimization schemes will be suggested to maximize $\|\Delta H\|_2$ in Section 5.2. First, however, in the next section we compute the asymptotic and monotonic stability radii for the interval ILC (IILC) problem with a given gain matrix and nominal plant, using the interval concepts and stability conditions given above.

5.1 Stability Radius

In this section, the Schur stability radius that satisfies sufficient stability conditions is calculated using the discrete Lyapunov equation. Let us introduce the symbol $\langle \cdot \rangle$ to represent the bigger norm value between a matrix and its transpose:

$$\langle \Delta T \rangle \equiv \max \left\{ \|(\Delta T)^T\|, \|\Delta T\| \right\}, \qquad (5.3)$$

where $\| \cdot \|$ denotes any kind of matrix norm. With this notation, the following theorem can be developed:

Theorem 5.2. *Given* Γ *designed for the nominal plant* H^o, *if there exists a symmetric, positive definite matrix* P *(i.e.,* $P = P^T > 0$*) that satisfies the constraint*

$$(I - H^o \Gamma)^T P (I - H^o \Gamma) - P = -I, \tag{5.4}$$

then the maximum allowable interval uncertainty (AIU) for which $(I - (H^o \pm \Delta H) \Gamma)$ is guaranteed to be AS, is bigger than $\langle \Delta H \rangle$ that satisfies

$$\langle \Delta H \rangle \equiv \frac{-\langle I - H^o \Gamma \rangle + \sqrt{\langle I - H^o \Gamma \rangle^2 + \frac{1}{\|P\|}}}{\langle \Gamma \rangle}. \tag{5.5}$$

■

Proof. Let us assume that $(I - H^o \Gamma)$ is Schur stable, so T^o is Schur stable. Then there exists $P = P^T > 0$ such that

$$(T^o)^T P T^o - P = -I. \tag{5.6}$$

If the following inequality is true with $P = P^T > 0$ from (5.6):

$$(T)^T P T - P < 0, \ T \in T^I, \tag{5.7}$$

then all $T \in T^I$ are Schur. Using $T = T^o - \Delta T$, (5.7) is changed to

$$(T^o - \Delta T)^T P (T^o - \Delta T) - P < 0$$
$$\Leftrightarrow (T^o)^T P T^o - (T^o)^T P \Delta T - (\Delta T)^T P T^o + (\Delta T)^T P \Delta T - P < 0. \tag{5.8}$$

Substituting (5.6) into (5.8), we have:

$$-(T^o)^T P \Delta T - (\Delta T)^T P T^o + (\Delta T)^T P \Delta T - I < 0$$
$$\Leftrightarrow -(T^o)^T P \Delta T - (\Delta T)^T P T^o + (\Delta T)^T P \Delta T < I, \tag{5.9}$$

where the left-hand side is a symmetric matrix. Therefore, if (5.9) is satisfied with $P = P^T$ determined by (5.6), then $T \in T^I$ is Schur stable. Now, taking the matrix norm of both sides of (5.9), we have

$$\| -(T^o)^T P \Delta T - (\Delta T)^T P T^o + (\Delta T)^T P \Delta T \| < 1. \tag{5.10}$$

Recall that if (5.10) is true, then (5.9) is true; not vice-versa (see Remark 5.3 for more explanation). Now we change (5.10) to

$$\|(T^o)^T \| \| P \| \| \Delta T \| + \|(\Delta T)^T \| \| P \| \| T^o \| + \|(\Delta T)^T \| \| P \| \| \Delta T \| < 1. \tag{5.11}$$

Notice that (5.11) is a sufficient condition for (5.10). Using $\Delta T = \Delta H \Gamma$ and the $\langle \cdot \rangle$ operator, the above inequality is changed to

$$\langle T^o \rangle \| P \| \langle \Delta T \rangle + \langle \Delta T \rangle \| P \| \langle T^o \rangle + \langle \Delta T \rangle \| P \| \langle \Delta T \rangle < 1$$
$$\Leftrightarrow [2 \langle \Delta T \rangle \langle T^o \rangle + \langle \Delta T \rangle^2] \| P \| < 1$$
$$\Leftrightarrow [2 \langle \Delta H \rangle \langle \Gamma \rangle \langle T^o \rangle + \langle \Delta H \rangle^2 \langle \Gamma \rangle^2] \| P \| < 1. \tag{5.12}$$

Also, notice that (5.12) is a sufficient condition for (5.11). Let $\alpha \equiv \langle \Gamma \rangle^2$; and $\beta \equiv \langle \Gamma \rangle \langle T^o \rangle$; and $x \equiv \langle \Delta H \rangle$. Then, (5.12) is of the form:

$$(2\beta x + \alpha x^2)\|P\| < 1 \Rightarrow \alpha x^2 + 2\beta x < \tfrac{1}{\|P\|}. \tag{5.13}$$

Here, since $\alpha > 0$,

$$\frac{-\beta - \sqrt{\beta^2 + \tfrac{\alpha}{\|P\|}}}{\alpha} < x < \frac{-\beta + \sqrt{\beta^2 + \tfrac{\alpha}{\|P\|}}}{\alpha}.$$

Using $x > 0$,

$$x < \frac{-\beta + \sqrt{\beta^2 + \tfrac{\alpha}{\|P\|}}}{\alpha}.$$

Therefore, the following inequality is satisfied:

$$\langle \Delta H \rangle < \frac{-\langle \Gamma \rangle \langle T^o \rangle + \sqrt{[\langle \Gamma \rangle \langle T^o \rangle]^2 + \tfrac{\langle \Gamma \rangle^2}{\|P\|}}}{\langle \Gamma \rangle^2}. \tag{5.14}$$

Using $\langle T \rangle = \langle I - H\Gamma \rangle$, (5.14) becomes

$$\langle \Delta H \rangle < \frac{-\langle I - H^o \Gamma \rangle + \sqrt{\langle I - H^o \Gamma \rangle^2 + \tfrac{1}{\|P\|}}}{\langle \Gamma \rangle}. \tag{5.15}$$

Finally, if there exists $P = P^T$ such that (5.6) is satisfied and inequality (5.15) is true, then T^I is Schur (but still, vice-versa is not true). Therefore, the maximum interval uncertainty is allowed to be more than $\langle \Delta H \rangle_{\max}$, which is defined as

$$\langle \Delta H \rangle_{\max} \equiv \frac{-\langle I - H^o \Gamma \rangle + \sqrt{\langle I - H^o \Gamma \rangle^2 + \tfrac{1}{\|P\|}}}{\langle \Gamma \rangle}. \tag{5.16}$$

∎

Remark 5.3. Let us understand why (5.10) is a sufficient condition for (5.9). Consider the following two inequalities for demonstration purposes: $A < I$ and $\|A\| < 1$, where A is a symmetric matrix. Note that $A < I$ can be written as $0 < I - A$. Then, since $I - A$ is a positive definite matrix, the eigenvalues of $I - A$ are all positive by Theorem 3.7 of [52]. However, because $\lambda(I - A) = 1 - \lambda(A)$ if and only if $\lambda(A) < 1$, then again we have $I - A$ is positive definite. Now consider $\|A\| < 1$. From the fact that A is symmetric and $\rho(A) \leq \|A\|$, if $\|A\| < 1$, then $\rho(A) < 1 \Leftrightarrow \max\{|\lambda(A)|\} < 1$. Therefore, if $\|A\| < 1$ is true, then $A < I$. However, vice-versa is not true. Thus, (5.10) is a sufficient condition for (5.9).

From Theorem 5.2, the following corollary is immediate:

Corollary 5.4. *If* $\Gamma = (H^o)^{-1}$, *the maximum AIU of the IILC is* $\tfrac{1}{\|\Gamma\|} = \tfrac{1}{\|(H^o)^{-1}\|}$.

∎

Proof. When $\Gamma = (H^o)^{-1}$, $I - H^o\Gamma = 0$; and from (5.6), since T is zero, P is equal to I. Also, since $\|\Gamma\| = \|\Gamma^T\|$, from (5.16), the maximum AIU is

$$\|\Delta H\| = \frac{1}{\|\Gamma\|} = \frac{1}{\|(H^o)^{-1}\|}. \tag{5.17}$$

∎

Now, let us consider the MC condition. For this we need the following definitions:

Definition 5.5. *The following augmented matrices are used for the MC analysis:*

$$T_s = \begin{bmatrix} 0 & (I - H\Gamma)^T \\ (I - H\Gamma) & 0 \end{bmatrix}, \ H \in H^I$$

$$T_s^o = \begin{bmatrix} 0 & (I - H^o\Gamma)^T \\ (I - H^o\Gamma) & 0 \end{bmatrix}.$$

∎

Also let $\|\cdot\|_2$ be the matrix 2-norm; $\langle\Gamma\rangle_k \equiv \max\{\|\Gamma\|_k, \|\Gamma^T\|_k\}$ and $\langle\Delta H\rangle_k \equiv \max\{\|\Delta H\|_k, \|\Delta H^T\|_k\}$. Then we have the following result.

Theorem 5.6. *Given Γ designed for the nominal plant H^o, if there exists a symmetric, positive definite matrix P_s that satisfies the constraint*

$$(T_s^o)^T P_s T_s^o - P_s = -I_{2n \times 2n}, \tag{5.18}$$

then the maximum allowable interval uncertainty (AIU) for which $(I - (H^o \pm \Delta H)\Gamma)$ is guaranteed to have MC, is bigger than $\langle\Delta H\rangle$ that satisfies

$$\langle\Delta H\rangle_k \equiv \frac{-\|T_s^o\|_2 + \sqrt{\|T_s^o\|_2^2 + \frac{1}{\|P_s\|_2}}}{\langle\Gamma\rangle_k}, \tag{5.19}$$

where k is 1 or ∞.

∎

Proof. The ILC system is given as

$$E_{k+1} = (I - H\Gamma)E_k, \ H \in H^I. \tag{5.20}$$

In Theorem 5.2, the condition for guaranteeing $\rho(I - H\Gamma) < 1$ (i.e., spectral radius less than 1) using the discrete Lyapunov inequality was found. The maximum singular value is defined as

$$\bar{\sigma}(I - H\Gamma) = \sqrt{\rho[(I - H\Gamma)^T(I - H\Gamma)]}. \tag{5.21}$$

Thus, the following relationship is true:

$$[\bar{\sigma}(I - H\Gamma)]^2 = \bar{\lambda}\begin{bmatrix} 0_{n \times n} & (I - H\Gamma)^T \\ (I - H\Gamma) & 0_{n \times n} \end{bmatrix} = \bar{\lambda}(T_s), \tag{5.22}$$

where $\bar{\lambda}$ is the maximum eigenvalue. Therefore, since the maximum eigenvalue of the right-hand side equals the spectral radius, if $\rho[(I - H\Gamma)^T(I - H\Gamma)] < 1$, then $\bar{\sigma}(I - H\Gamma) < 1$. Thus, the singular value stability problem is changed to an eigenvalue problem by Definition 5.5. Since the eigenvalues of $(I - H\Gamma)^T(I - H\Gamma)$ are equal to the eigenvalues of T_s, the discrete Lyapunov inequality can be applied to T_s. If T_s^o is Schur stable, then the following is true:

$$(T_s^o)^T P_s T_s^o - P_s = -I_{2n \times 2n}, \tag{5.23}$$

with $P_s = P_s^T > 0$. If T_s, $T_s \in T_s^I$ is stable, the following is also true:

$$(T_s)^T P_s T_s - P_s < 0. \tag{5.24}$$

Thus the following relationships can be derived:

$$
\begin{aligned}
T_s^o - T_s &= \Delta T_s \\
&= \begin{bmatrix} 0_{n \times n} & (I - H^o\Gamma - (I - H\Gamma))^T \\ (I - H^o\Gamma - (I - H\Gamma)) & 0_{n \times n} \end{bmatrix} \\
&= \begin{bmatrix} 0_{n \times n} (\Delta H\Gamma)^T \\ (\Delta H\Gamma) 0_{n \times n} \end{bmatrix}.
\end{aligned} \tag{5.25}
$$

Note, T_s^o and ΔT_s are symmetric matrices. Thus, $\|T_s^o\|_2 = \|(T_s^o)^T\|_2$ and $\|\Delta T_s\|_2 = \|\Delta T_s^T\|_2$.

Now, let us change (5.24) to be:

$$(T_s^o - \Delta T_s)^T P_s (T_s^o - \Delta T_s) - P_s < 0$$
$$\Leftrightarrow (T_s^o)^T P_s T_s^o - (T_s^o)^T P_s \Delta T_s - (\Delta T_s)^T P_s T_s^o + (\Delta T_s)^T P_s \Delta T_s - P_s < 0. \tag{5.26}$$

Using (5.23), the above inequality is changed to

$$-(T_s^o)^T P_s \Delta T_s - (\Delta T_s)^T P_s T_s^o + (\Delta T_s)^T P_s \Delta T_s < I_{2n \times 2n}, \tag{5.27}$$

and taking the 2-norm of both sides, we get

$$\left[2\|\Delta T_s\|_2 \|T_s^o\|_2 + \|\Delta T_s\|_2^2 \right] \|P_s\|_2 < 1. \tag{5.28}$$

Here, it is necessary to separate $\|\Delta T_s\|_2$ into $\|\Delta H\|$ and $\|\Gamma\|$. For this purpose, the following inequality is used:

$$\|\Delta T_s\|_2 \leq \sqrt{\|\Delta T_s\|_1 \|\Delta T_s\|_\infty}. \tag{5.29}$$

Thus, from the following relationship:

$$
\begin{aligned}
\|\Delta T_s\|_1 &= \left\| \begin{bmatrix} 0_{n \times n} & (\Delta H\Gamma)^T \\ (\Delta H\Gamma) & 0_{n \times n} \end{bmatrix} \right\|_1 \\
&= \max\{\|\Delta H\Gamma\|_1, \|(\Delta H\Gamma)^T\|_1\}
\end{aligned} \tag{5.30}
$$

$$\|\Delta T_s\|_\infty = \left\|\begin{bmatrix} 0_{n\times n} & (\Delta H\Gamma)^T \\ (\Delta H\Gamma) & 0_{n\times n} \end{bmatrix}\right\|_\infty$$
$$= \max\{\|\Delta H\Gamma\|_\infty, \|(\Delta H\Gamma)^T\|_\infty\}, \qquad (5.31)$$

and using the $\langle\cdot\rangle$ operator, we have

$$\langle\Delta H\Gamma\rangle = \max\{\|\Delta H\Gamma\|_1, \|(\Delta H\Gamma)^T\|_1\} = \max\{\|\Delta H\Gamma\|_\infty, \|(\Delta H\Gamma)^T\|_\infty\}. \qquad (5.32)$$

However, notice that in $\langle\Delta H\Gamma\rangle$ only the 1- and ∞-norms are effective. Thus, (5.29) is changed as

$$\|\Delta T_s\|_2 \leq \sqrt{\|\Delta T_s\|_1 \|\Delta T_s\|_\infty}$$
$$= \sqrt{\langle\Delta H\Gamma\rangle\langle\Delta H\Gamma\rangle}$$
$$= \langle\Delta H\Gamma\rangle$$
$$\leq \langle\Delta H\rangle\langle\Gamma\rangle. \qquad (5.33)$$

Now, let us substitute $\langle\Delta H\rangle\langle\Gamma\rangle$ into $\|\Delta T_s\|_2$ of (5.28) to obtain the following sufficient inequality:

$$\left[2\langle\Delta H\rangle\langle\Gamma\rangle\|T_s^o\|_2 + (\langle\Delta H\rangle\langle\Gamma\rangle)^2\right] \|P_s\|_2 < 1. \qquad (5.34)$$

Then, using the same procedure as in the proof of Theorem 5.2, we have

$$\langle\Delta H\rangle < \frac{-\|T_s^o\|_2 + \sqrt{\|T_s^o\|_2^2 + \frac{1}{\|P_s\|_2}}}{\langle\Gamma\rangle}.$$

Therefore, the maximum allowable interval uncertainty is calculated as

$$\langle\Delta H\rangle_{\max} = \frac{-\|T_s^o\|_2 + \sqrt{\|T_s^o\|_2^2 + \frac{1}{\|P_s\|_2}}}{\langle\Gamma\rangle}, \qquad (5.35)$$

where $\langle\Delta H\rangle_{\max}$ and $\langle\Gamma\rangle$ are restricted to the 1- and ∞-norms. ∎

In Theorem 5.6, the AIU, $\langle\Delta H\rangle_{\max}$, is calculated in 1- or ∞-norms. The AIU can be calculated for the 2-norm, using the following relationship:

$$\langle\Gamma\rangle_1 = \max\{\|\Gamma\|_1, \|\Gamma^T\|_1\} = \max\{\|\Gamma\|_\infty, \|\Gamma^T\|_\infty\} = \langle\Gamma\rangle_\infty. \qquad (5.36)$$

Then we have the following corollary:

Corollary 5.7. *The 2-norm-based maximum AIU is greater than:*

$$\langle\Delta H\rangle_2 \equiv \frac{-\|T_s^o\|_2 + \sqrt{\|T_s^o\|_2^2 + \frac{1}{\|P_s\|_2}}}{\langle\Gamma\rangle_k}, \qquad (5.37)$$

where $k = 1$ or ∞. Notice that the right-hand side of (5.37) is equivalent to the right-hand side of (5.35). ∎

Proof. By Theorem 5.6 and (5.36), the following is true:

$$\langle \Delta H \rangle_1 = \langle \Delta H \rangle_\infty.$$

Also, the following inequality is satisfied:

$$\langle \Delta H \rangle_2 \leq \sqrt{\langle \Delta H \rangle_1 \langle \Delta H \rangle_\infty}$$

and the following relationship is immediate:

$$\langle \Delta H \rangle_2 \leq \frac{-\|T_s^o\|_2 + \sqrt{\|T_s^o\|_2^2 + \frac{1}{\|P_s\|_2}}}{\langle \Gamma \rangle_k}. \tag{5.38}$$

■

For convenience, let us denote the right-hand side of (5.16) as ΔH_{asym} and the right-hand side of (5.37) as ΔH_{mono}. Then, it is concluded that if the interval uncertainty in ILC is less than ΔH_{asym}, the ILC system is asymptotically stable and, if the interval uncertainty in ILC is less than ΔH_{mono}, the ILC process is MC in the l_2-norm topology of E_k.

Remark 5.8. Theorem 5.2 and Corollary 5.4 are satisfied with any kind of norms. In Theorem 5.6, $\langle \Delta H \rangle$ is the 1-norm or the ∞-norm, and in Corollary 5.7, $\langle \Delta H \rangle$ is 2-norm. In Theorem 5.6 and Corollary 5.7, $\|T_s^o\|$ and $\|P_s\|$ are 2-norms, and $\langle \Gamma \rangle$ is the 1- or ∞-norm.

Remark 5.9. In Definition 5.1, the maximum Schur stability radius was defined. ΔH_{asym} and ΔH_{mono} provide sufficient stability radii for the IILC. "Sufficient" means that the actual maximum stability radii, ΔH_a^r and ΔH_m^r, may be bigger than the calculated stability radius, ΔH_{asym} and ΔH_{mono}, respectively. Thus, the following inequalities should be noted:

$$\Delta H_{asym} \leq \Delta H_a^r; \ \Delta H_{mono} \leq \Delta H_m^r.$$

Hence, ΔH_{asym} and ΔH_{mono} will be conservative compared with the actual maximum stability radii.

So far, we have found maximum AIU bounds for both AS and MC given Γ designed for H^o. In the next section, optimization methods are used to design Γ in order to maximize the stability radius for a given H^o.

5.2 Optimization

In this section, two optimization schemes are suggested based on Section 5.1. The purpose of the optimization is to maximize ΔH_{asym} and ΔH_{mono} by designing Γ, with the constraint that the IILC system is either asymptotically stable or monotonically-convergent. To find the optimal Γ that allows more interval uncertainties in terms of AS, the following optimization scheme is suggested:

$$\max_{\Gamma} \Delta H_{asym} \tag{5.39}$$

$$\text{s.t. } (I - H^o \Gamma)^T P(I - H^o \Gamma) - P = -I. \tag{5.40}$$

The same optimization idea can be used for increasing the uncertainty interval of the system in terms of MC:

$$\max_{\Gamma} \Delta H_{mono} \tag{5.41}$$

$$\text{s.t. } (T_s^o)^T P_s T_s^o - P_s = -I_{2n \times 2n}. \tag{5.42}$$

Remark 5.10. It is easy to see that the maximum interval uncertainties occur when $\Gamma = 0$, because when $\Gamma = 0$, there could be infinity interval uncertainties in H^o. From $I - H\Gamma$, $H \in H^I$ it is easy to observe that as $\Gamma \to 0$, even though $H \to \infty$, the following is true: $\|I - H\Gamma\| < 1$. Thus, in the optimization problem, the required maximum spectral radius and singular value should be fixed. In other words, we should add one more constraint in the optimization schemes, either $\rho(I - H^o \Gamma) < \rho_{max}$ or $\overline{\sigma}(I - H^o \Gamma) < \overline{\sigma}_{max}$, where $\rho_{max} < 1$ and $\overline{\sigma}_{max} < 1$.

Remark 5.11. In the above two optimization schemes, ρ_{max} and $\overline{\sigma}_{max}$ are design parameters. If these values are near zero, the system converges quickly, but with the trade-off that there could be a small AIU. On the other hand, if these values are near one, the system converges slowly, but allows a large AIU. Thus we must choose ρ_{max} and $\overline{\sigma}_{max}$ before applying the optimization schemes.

5.3 Simulation Illustrations

Several simulation results are presented in this section to demonstrate the ideas developed in this chapter. The following discrete system is considered:

$$x_{k+1} = \begin{bmatrix} -0.50 & 0.00 & 0.00 \\ 1.00 & 1.24 & -0.87 \\ 0.00 & 0.87 & 0.00 \end{bmatrix} x_k + \begin{bmatrix} 1.0 \\ 0.0 \\ 0.0 \end{bmatrix} u_k \tag{5.43}$$

$$y_k = \begin{bmatrix} 2.0 & 2.6 & -2.8 \end{bmatrix} x_k. \tag{5.44}$$

The system has poles at $[0.62 + j0.62, 0.62 - j0.62, -0.50]$ and zeros at $[0.65, -0.71]$. The main hypothesis of this simulation test is initial reset. In other words, it is assumed that the system starts at the same place at every iteration. Another hypothesis is interval model uncertainty. That is, the model uncertainty associated with the nominal Markov plant is bounded by two extreme boundary matrices.

5.3.1 Test Setup

The simulation test is performed with the following reference sinusoidal signal: $Y_d = \sin(8.0j/n)$, where $n = 10$ and $j = 1, \ldots, n$. The band size, which is defined as the number of Arimoto-like, causal and noncausal bands used in Γ (see Section 4.1 for the definition of "limited-band size"), is fixed at 3 (i.e., the Arimoto-like band, one causal band, and one noncausal band) and learning gains are determined by optimization problems described in Section 5.2. Since the gains of each band are not fixed at the same value, the ILC algorithm is considered to be linear, time-varying, and noncausal. The uniformly distributed random number generator of MATLAB® was used to make interval uncertainties in Markov parameters according to: $h_i = h_i + \delta|h_i|w$, where $w \in [-1, 1]$ is a uniformly distributed random number; and δ is tuned to limit the interval amount (in the matrix 2-norm). First the optimal learning gain matrices are designed from the optimization problems suggested in Section 5.2. In MATLAB®, the nonlinear optimization command $fmincon$ was used to solve these problems. Then, using the resulting learning gain matrix, an ILC experiment was performed with each of 1000 different random plants. For each random plant, 20 iterations were carried out as shown in Table 5.1. The design parameters ρ_{max} and $\bar{\sigma}_{max}$ were selected as 0.9. The monotonic convergence optimization scheme was designed assuming an l_2-norm topology for E_k. Thus, if there exists an optimization solution for ΔH_{mono}, then $(\sum_{i=1}^{n} |E_k(i)|^2)^{1/2}$ will be MC.

Table 5.1. Simulation Setup for Monte-Carlo-Type Random Test

for $i = 1 : 1 : 1000$
Pick a random plant
for $j = 1 : 1 : 20$
Repeat iterative test
end
end

5.3.2 Test Results

From the optimization problems of (5.39) and (5.41), the learning gain matrices were designed using the nominal plant such that the calculated maximum AIUs become $\Delta H_{asym} = 0.737$, and $\Delta H_{mono} = 0.6954$. The physical meaning of $\Delta H_{asym} = 0.737$ is that the ILC gain matrix designed from optimization (5.39) allows interval uncertainty for the nominal Markov matrix an amount $\|\Delta H\|_2 < 0.737$ while ensuring AS. The physical meaning of $\Delta H_{mono} = 0.6954$ is that the ILC gain matrix designed from optimization (5.41) allows interval uncertainty for the nominal Markov matrix by an amount $\|\Delta H\|_2 < 0.6954$ while still ensuring MC.

The results are illustrated in Figure 5.1 for the AS case and Figure 5.2 for the MC case. In Figure 5.1 tests were performed using the ILC learning gain matrix designed from optimization (5.39); the left-hand figures show the interval amount of random plants in matrix 2-norms; the right-hand figures show ILC performance corresponding to the left-hand figures; circle-marked lines are the l_2-norm errors vs. ILC iteration number corresponding to the plant with the maximum matrix 2-norm and diamond-marked lines are l_2-norm errors vs. ILC iteration number corresponding to the plant with the minimum matrix 2-norm. In Figure 5.2, tests were performed using the ILC learning gain matrix designed from optimization (5.41); the left-hand figures show the interval amount of random plants in matrix 2-norms; the right-hand figures show ILC performance corresponding to the left-hand figures; circle-marked lines are l_2-norm errors vs. ILC iteration number corresponding to the plant with the maximum matrix 2-norm and diamond marked-lines are l_2-norm errors vs. ILC iteration number corresponding to the plant with the minimum matrix 2-norm. In both figures, the left-hand side of the figures shows plots of the 2-norm of the various random plants used (the index of the 1000 different plants is shown on the horizontal axis with the resulting 2-norm of the plant given on the vertical axis). The right-hand side of the figures shows the maximum and minimum l_2 norm of the super-vector error plotted as a function of iteration.

First, let us check the validity of $\Delta H_{asym} = 0.737$. To check the validity of this value, we gave random intervals to each Markov parameter, and we selected interval plants with $\|\Delta H\|_2$ less than 0.737. The results shown in Figure 5.1(a) meet the expectation that all these plants should converge asymptotically. However, as commented in Remark 5.9, there could exist interval plants with $\|\Delta H\|_2 > 0.737$ that are asymptotically stable with the ILC gain matrix designed from (5.39), because the result is only sufficient. Figure 5.1(c) and Figure 5.1(d) show such a situation. But, as the perturbation grows beyond the bound ΔH_{asym} eventually we encounter plants for which the designed learning gain no longer gives AS. This is shown in Figure 5.1(e) and Figure 5.1(f).

Similarly, we can check the validity of $\Delta H_{mono} = 0.6954$. Figure 5.2(a) and Figure 5.2(b) show the situation when $\|\Delta H\|_2 < \Delta H_{mono} = 0.6954$. We see that the ILC gain matrix designed from (5.41) guarantees the MC. As in the asymptotic stability (AS) example, the remaining plots in Figure 5.2 show the sufficiency of the condition and the final instability.

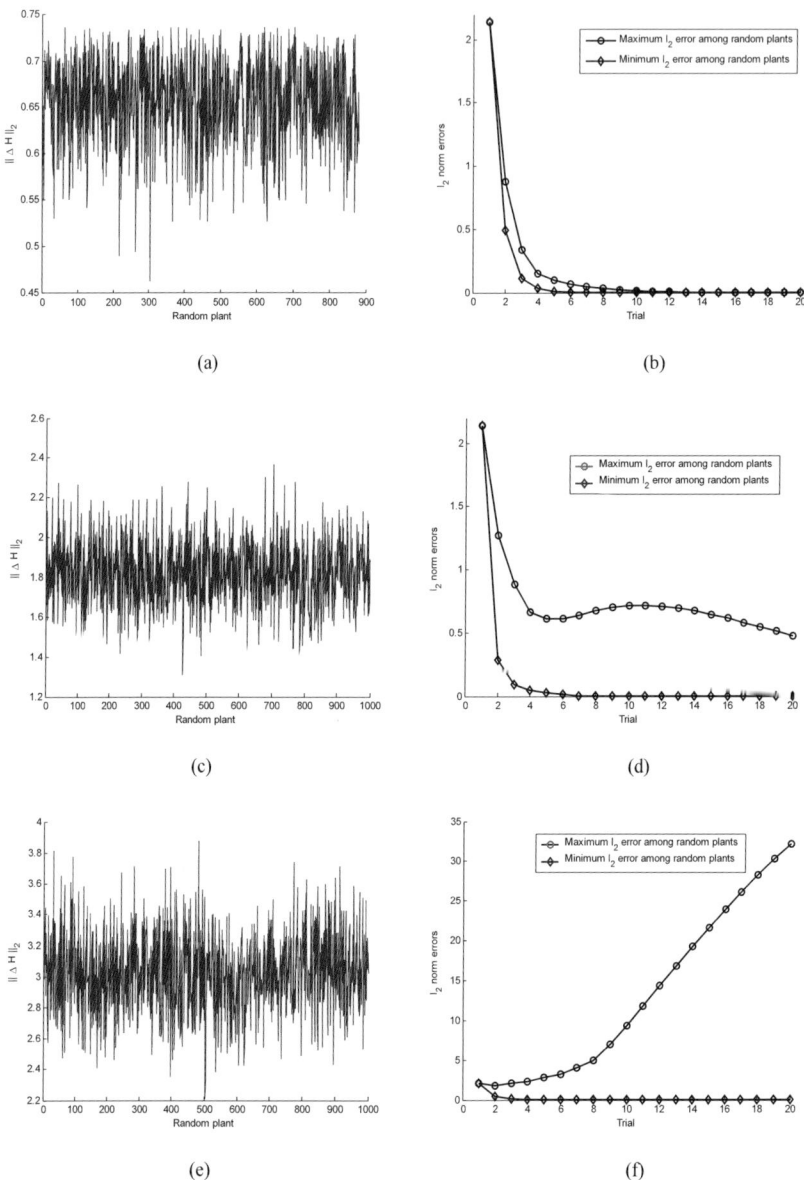

Fig. 5.1. Asymptotic stability tests with maximized stability radius. Left-hand figures show the interval amount of random plants in matrix 2-norms. Right-hand figures show ILC performance corresponding to left-hand figures. Circle-marked lines are the l_2-norm errors vs. ILC iteration number corresponding to the plant with the maximum matrix 2-norm and diamond-marked lines are l_2-norm errors vs. ILC iteration number corresponding to the plant with the minimum matrix 2-norm. Panel (b) shows that IILC is MC with interval uncertainties given in Panel (a), and Panel (d) shows that IILC is AS with interval uncertainties given in Panel (c). However, Panel (f) shows that IILC is divergent with interval uncertainties given in Panel (e).

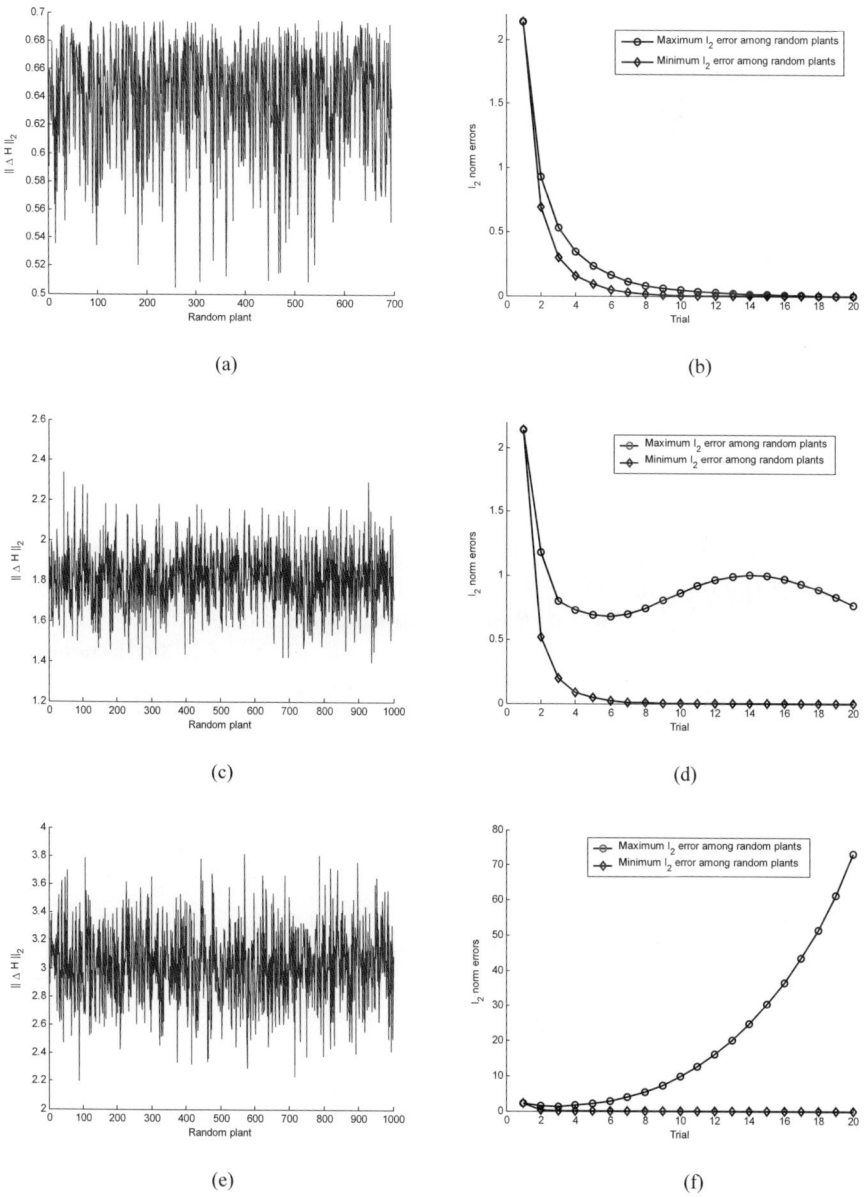

Fig. 5.2. Monotonic convergence tests with maximized stability radius. Left-hand figures show the interval amount of random plants in matrix 2-norms. Right-hand figures show ILC performance corresponding to left-hand figures. Circle-marked lines are the l_2-norm errors vs. ILC iteration number corresponding to the plant with the maximum matrix 2-norm and diamond-marked lines are l_2-norm errors vs. ILC iteration number corresponding to the plant with the minimum matrix 2-norm. Panel (b) shows that IILC is MC with interval uncertainties given in Panel (a), and Panel (d) shows that IILC is AS with interval uncertainties given in Panel (c). However, Panel (f) shows that IILC is divergent with interval uncertainties given in Panel (e).

5.4 Chapter Summary

In this chapter, bounds on the maximum allowable uncertainty in the plant Markov parameters for both AS and MC were calculated. These bounds were then used to design the ILC learning gain matrix to maximize the asymptotic and monotonic stability radii of the nominal plant. Simulation results illustrated the ideas. This approach provides an effective scheme for designing a robust ILC system under interval plant uncertainty. As a main development of this chapter, a discrete Lyapunov equation is used to compute the Schur stability radius of the ILC system for the case of parametric or interval perturbations in the system Markov parameters. We also showed how to design the ILC learning gain matrix such that the system can be asymptotically stable and/or monotonically-convergent for the worst case uncertainty. The proposed approach allows design of a causal or noncausal time-varying learning gain matrix.

The main contribution of this chapter is to introduce well-established concepts from interval/parametric robust control theory [314] into the ILC paradigm, with a distinct focus on the iteration domain. Specific contributions are the introduction of the idea of interval uncertainty and design techniques to deal with such uncertainty. Note that while other works in the ILC literature have considered robustness (e.g., [60, 63, 61, 64, 351]), none has considered interval robustness and none has considered this robustness in the iteration domain. Thus, this chapter makes unique (iteration-domain focus) and relevant (interval uncertainty focus fills in a gap in the existing literature) contributions to the ILC literature.

6

Iterative Learning Control Design Based on Interval Model Conversion

Although the practical importance of the monotonic convergence issue (MC) has been well-addressed in the ILC literature (see [279] for a nice discussion of this problem), to date this issue has not been fully understood with respect to model uncertainty. In this chapter we consider the problem in more detail by presenting another ILC algorithm using causal/noncausal and Arimoto-like band-limited, time-varying learning gains for designing a monotonically-convergent ILC process when the plant is subject to parametric uncertainties. Our strategy is to use use first-order perturbation theory to find bounds on the eigenvalues and eigenvectors of the powers of A when A is an interval matrix. These bounds are then used for calculation of the interval uncertainty of the Markov matrix H. Next the bounds on the Markov matrix are used to design an iterative learning controller that ensures MC for all systems in the interval plant via the solution of an optimization problem (actually, two optimization problems are suggested). Note the difference between the design results in this chapter and the design results of the previous chapter. In the previous chapter, optimization was used to find a learning gain Γ so that the ILC system could tolerate the largest possible uncertainty. Here we assume the uncertainty bounds are given or computed (i.e., $\overline{H}$ and $\underline{H}$) and we seek a Γ to ensure MC for all plants defined by those uncertainty bounds.

6.1 Interval Model Conversion in ILC

In the super-vector framework, Markov parameters are used for ILC design. To accomplish the robust ILC design that we seek to demonstrate in this chapter, the interval uncertainty of the nominal plant, either using state-space or transfer function models, has to be converted into interval uncertainties in the Markov parameters. We call this process "interval model conversion." Let us focus on the nominal discrete-time system model given in the following SISO state-space form:

$$x(t+1) = Ax(t) + Bu(t); \ y(t) = Cx(t), \tag{6.1}$$

where $A, B,$ and C are the matrices describing the system in the state-space and $x(t), u(t),$ and $y(t)$ are the state, input, and output variables, respectively. Without loss of generality, the relative degree of the system is assumed to be 1, so $CB \neq 0$. The Markov parameters of the plant are defined as $h_k = CA^{k-1}B$. These Markov parameters can then be used to form the Markov matrix, a lower-triangular Toeplitz matrix whose first column is the vector $[h_1, h_2, \ldots, h_N]^T$ and that is denoted as H (see (1.27)). Then, the interval conversion process is to find bounds on h_k from the bounds on uncertain A, $B,$ and C using $h_k = CA^{k-1}B$. To simplify the presentation, it is assumed that model uncertainty exists in A only. It is easy to extend the approach to uncertain B and C matrices as well. But, since the key issue in interval model conversion is to find the power of an interval A matrix (as commented in [244], it is NP-hard to find exact boundaries of the power of an interval matrix system), this monograph only focuses on systems with an uncertain A-matrix. Then, *interval model conversion* can be explicitly defined as the process of finding the uncertain boundaries of $h_k \in h_k^I = [\underline{h_k}, \overline{h_k}]$ from A^I, B and C.

To clarify the interval perturbation concepts, we introduce the following definition.

Definition 6.1. *Let $A \in A^I$ be a member of the interval matrix $A^I = A^o + \Delta A^I$, where $A^o = [a_{ij}^o]$ is the nominal plant matrix and ΔA^I is the interval perturbation matrix defined as*

$$\Delta A^I := \left\{ \Delta A = \left[\Delta a_{ij} : \Delta a_{ij} \in [\underline{a_{ij}} - a_{ij}^o, \overline{a_{ij}} - a_{ij}^o] \right] \right\}. \tag{6.2}$$

∎

Next, to proceed we make two assumptions. The first is a technical assumption and the second is more practical, as will be described in a later remark.

Assumption 6.2. *The nominal plant matrix A^o is in the interval matrix A^I, and each uncertain fixed matrix A is an element of the interval matrix A^I. Furthermore, each uncertain matrix is diagonalizable so that it can be decomposed as $A = X\Lambda X^{-1}$, with $\Lambda = \mathrm{diag}(\lambda_i)$, where λ_i are the eigenvalues of A.*

Assumption 6.3. *Every matrix $A \in A^I$ is Schur stable.*

Using these assumptions, the following theorem is established to estimate the boundary of the power of interval matrix.

Theorem 6.4. *Let A^I be an $n \times n$ interval matrix and let Λ^I, X^I, Y^I be $n \times n$ interval matrices such that for each $A \in A^I$ there exist a diagonal matrix $\Lambda \in \Lambda^I$ and a matrix $X \in X^I$ such that $A = X\Lambda X^{-1}$ holds and $X^{-1} \in Y^I$. Then for each $k \geq 1$ we have*

$$\left\{ A^k : A \in A^I \right\} \subseteq X^I \otimes (\Lambda^I)^k \otimes Y^I, \tag{6.3}$$

where $\otimes$ *(including the power) is the interval arithmetic multiplication operator defined in Chapter 4 (also, refer to [314, 8, 215]).* ∎

Proof. The proof is straightforward since any $A \in A^I$ can be written, according to assumptions, as $A = X \Lambda X^{-1}$ for some $\Lambda \in \Lambda^I$ and $X \in X^I$ which additionally satisfies $X^{-1} \in Y^{-1}$. Then $A^k = X \Lambda^k X^{-1} \in X^I \otimes (\Lambda^I)^k \otimes Y^I$ due to the basic "inclusion isotony" of interval arithmetic operations (see e.g. [215]), which proves the inclusion (6.3). ∎

From now on, for convenience, let us use the notation

$$(A^I)^k := \left\{ A^k : A \in A^I \right\} \text{ and } A_k^I := X^I \otimes (\Lambda^I)^k \otimes Y^I, \tag{6.4}$$

which means that $(A^I)^k \subseteq A_k^I$. From a computational perspective, the boundary of $(A^I)^k$ can be estimated by multiplying the interval matrix using interval calculation software such as Intlab [179], but the result will be quite conservative as k increases and it requires a huge amount of computation. However, since $(A^I)^k \subseteq A_k^I$, the boundary of $(A^I)^k$ can be estimated by estimating the boundary of A_k^I, which is also easily done in Intlab, using the boundaries of the three interval matrices Λ^I, X^I, and Y^I, as explained in the following section. Observe also that if the maximum absolute eigenvalue of $A \in A^I$ is less than 1, for all $\mathcal{A}_k \in A_k^I$, $\mathcal{A}_k$ will converge to zero as $k \to \infty$ because Λ^k, $\Lambda \in \Lambda^I$, converges to zero as $k \to \infty$. However, if the maximum absolute eigenvalue is bigger than 1, $\mathcal{A}_k \in A_k^I$ could diverge. Then the bound on the uncertain interval boundary of h_k becomes bigger and bigger as k increases. For this reason it was assumed that $A \in A^I$ is Schur stable.

In this section, it was shown that we can contain the original interval system inside a "bigger" interval system according to: $(A^I)^k \subseteq A_k^I$. Therefore, the remaining work is to estimate the boundaries of Λ^I, X^I, and Y^I and from these estimates to compute bounds on A^k. The next section suggests an analytical method to estimate the bounds of Λ^I, X^I, and Y^I from A^I using first-order perturbation theory [291].

6.2 Interval Matrix Eigenpair Bounds

In this section, first-order perturbation theory is briefly summarized and then two lemmas are suggested to obtain analytical solutions of the boundaries of X^I, Λ^I and Y^I from A^I, to be used for estimating the boundary of A_k^I. Let us suppose that the $n \times n$ nominal matrix A^o has its n different nominal eigenvalues λ_{0i}, particular nominal left eigenvectors x_{0i}, and particular nominal right eigenvectors y_{0i}, where $i = 1, \dots, n$, defined by $x_{0i}^T A^o = \lambda_{i0} x_{0i}^T$ and $A^o y_{0i} = \lambda_{i0} y_{0i}$. Further define the nominal eigenvalue matrix, Λ^o, particular nominal left eigenvector matrix, X^o, and particular nominal right eigenvector matrix, Y^o, by $\Lambda^o = [\lambda_{ij} : \lambda_{ij} = \lambda_{0i} \text{ if } i = j, \ \lambda_{ij} = 0 \text{ if } i \neq j]$,

$X^o = [x_{01}, x_{02}, \ldots, x_{0n}]^T$, and $Y^o = [y_{01}, y_{02}, \ldots, y_{0n}]$, respectively, with $(Y^o)^{-1} = X^o$. Next, assume a perturbation to A^o to form $A = A^o + \Delta A$, where $A \in A^I$ and $\Delta A \in \Delta A^I$, based on Definition 4.5 and Definition 6.1. Denote the eigenvalues and left and right eigenvectors associated with the perturbation ΔA as λ_{1i}, x_{1i}, and y_{1i}, respectively. In other words, when λ_i, x_i, and y_i represent eigenvalues and eigenvectors of $A \in A^I$, the following relationships are satisfied: $\lambda_i = \lambda_{0i} + \lambda_{1i}$, $x_i = x_{0i} + x_{1i}$, and $y_i = y_{0i} + y_{1i}$.[1] Observe that the set of λ_{1i}, x_{1i}, and y_{1i} define scalar intervals. In this section, we are interested in finding the boundaries of λ_{1i}, x_{1i}, and y_{1i}, which we will use to estimate the boundaries of λ_i, x_i, and y_i.

From [291], the following formulae are adopted for the perturbed eigenvalues:

$$\lambda_{1i} = x_{0i}^T \Delta A y_{0i}, \ \forall \ \Delta A \in \Delta A^I \tag{6.5}$$

and for the perturbed eigenvectors we use:

$$x_{1i} = \sum_{k=1}^{n} \gamma_{ik} x_{0k}; \ y_{1i} = \sum_{k=1}^{n} \varepsilon_{ik} y_{0k}, \tag{6.6}$$

where $\gamma_{ik} = \frac{y_{0k}^T \Delta A x_{0i}}{\lambda_{0i} - \lambda_{0k}}$; $\varepsilon_{ik} = \frac{x_{0k}^T \Delta A y_{0i}}{\lambda_{0i} - \lambda_{0k}}$, $i, k = 1, \ldots, n; i \neq k$, and $\gamma_{ik} = \varepsilon_{ik} = 0$, $i = k$, $\forall \ \Delta A \subset \Delta A^I$ However, notice here that $\Delta A \in \Delta A^I$, where ΔA^I is an interval perturbation matrix, so it is quite messy to calculate λ_{1i}, x_{1i}, and y_{1i} in (6.5) and (6.6). For reliable analytical calculation of (6.5) and (6.6), the maximum absolute value of the real and imaginary part are considered separately. Let us denote the maximum of λ_{1i} on the real axis by $\lambda_{1i}|_{real}^{max}$, which can be defined as $\lambda_{1i}|_{real}^{max} = \max\{\lambda_{1i}|_{real} : \lambda_{1i}|_{real} = |\text{Re}(\lambda_{1i})|, \ \lambda_{1i} = x_{0i}^T \Delta A y_{0i}, \ \forall \ \Delta A \in \Delta A^I\}$, and the maximum of λ_{1i} on the imaginary axis by $\lambda_{1i}|_{imag}^{max}$, which can be defined as $\lambda_{1i}|_{imag}^{max} = \max\{\lambda_{1i}|_{imag} : \lambda_{1i}|_{imag} = |\text{Im}(\lambda_{1i})|, \ \lambda_{1i} = x_{0i}^T \Delta A y_{0i}, \ \forall \ \Delta A \in \Delta A^I\}$, where Re means the real part and Im means the imaginary part. Then, the following lemma is suggested.

Lemma 6.5. *Considering the real part and imaginary part separately, at any fixed i in the above definition (i.e., at a fixed eigenvalue), the following are true:*

$$\lambda_{1i}|_{real}^{max} = \max_{(\Delta A)^v \subset \Delta A^I} \{\text{Re}[x_{0i}^T (\Delta A)^v y_{0i}]\} \tag{6.7}$$

$$\lambda_{1i}|_{imag}^{max} = \max_{(\Delta A)^v \subset \Delta A^I} \{\text{Im}[x_{0i}^T (\Delta A)^v y_{0i}]\}, \tag{6.8}$$

where $(\Delta A)^v$ are the vertex matrices of ΔA^I. ∎

---

[1] Note that we do not imply that the eigenvalues of the sum of two matrices are equal to the sum of the eigenvalues of the individual matrices. Rather, the eigenpairs λ_{1i}, x_{1i}, and y_{1i} are "perturbation" eigenpairs and represent what would be added to the nominal values to obtain equivalent eignepairs of the perturbed matrix. Also note, on page 103 of [291], λ_{1i}, x_{1i}, and y_{1i} are called the first-order perturbation eigensolution.

Lemma 6.5 says we can compute the radius of the perturbed eigenvalues using only the vertex matrices of A. The proof of this result can be carried out using the same procedure shown below in the proof of Lemma 6.6. As the proof of Lemma 6.6 is more comprehensive, the proof of Lemma 6.5 is omitted.

Next, we show that the radii of the perturbed eigenvectors can also be estimated using the finite set of vertex matrices. First, let us consider the left eigenvectors and let us denote the jth element of x_{1i} by $(x_{1i})_j$. Then, denoting the maximum of $(x_{1i})_j$ on the real axis by $(x_{1i})_j|_{real}^{\max}$, which can be defined as

$$(x_{1i})_j|_{real}^{\max} = \max\{(x_{1i})_j|_{real} : (x_{1i})_j|_{real} = |\text{Re}((x_{1i})_j)|,$$

$$(x_{1i})_j = \sum_{k=1}^{n} \frac{y_{0k}^T \Delta A x_{0i}}{\lambda_{0i} - \lambda_{0k}} (x_{0k})_j, \ \forall \ \Delta A \in \Delta A^I\}, \quad (6.9)$$

and denoting the maximum of $(x_{1i})_j$ on the imaginary axis by $(x_{1i})_j|_{imag}^{\max}$, which can be defined as

$$(x_{1i})_j|_{imag}^{\max} = \max\{(x_{1i})_j|_{imag} : (x_{1i})_j|_{imag} = |\text{Re}((x_{1i})_j)|,$$

$$(x_{1i})_j = \sum_{k=1}^{n} \frac{y_{0k}^T \Delta A x_{0i}}{\lambda_{0i} - \lambda_{0k}} (x_{0k})_j, \forall \ \Delta A \in \Delta A^I\}, \quad (6.10)$$

we provide the following lemma (the perturbed radii of right eigenvectors are calculated in the same way).

Lemma 6.6. *Considering the real part and imaginary parts separately, at any fixed i and fixed j (i.e., at a fixed element of the fixed eigenvector), the maximum perturbation of $(x_{1i})_j$ can be calculated by checking the vertex matrices of the interval perturbation matrix according to:*

$$(x_{1i})_j|_{real}^{\max} = \max_{(\Delta A)^v \subset \Delta A^I} \left\{ \text{Re} \left[\sum_{k=1}^{n} \frac{y_{0k}^T (\Delta A)^v x_{0i}}{\lambda_{0i} - \lambda_{0k}} (x_{0k})_j \right] \right\} \quad (6.11)$$

$$(x_{1i})_j|_{imag}^{\max} = \max_{(\Delta A)^v \subset \Delta A^I} \left\{ \text{Im} \left[\sum_{k=1}^{n} \frac{y_{0k}^T (\Delta A)^v x_{0i}}{\lambda_{0i} - \lambda_{0k}} (x_{0k})_j \right] \right\}, \quad (6.12)$$

where $(x_{0k})_j$ is the jth element of the kth eigenvector x_{0k}. ∎

Proof. From (6.6), we have

$$x_{1i} = \gamma_{i1} x_{01} + \cdots + \gamma_{in} x_{0n} = \frac{y_{01}^T \Delta A x_{0i}}{\lambda_{0i} - \lambda_{01}} x_{01} + \cdots + \frac{y_{0n}^T \Delta A x_{0i}}{\lambda_{0i} - \lambda_{0n}} x_{0n}. (6.13)$$

In (6.13), since the denominators $\lambda_{0i} - \lambda_{01}, \ldots, \lambda_{0i} - \lambda_{0n}$ are scalars and y_{01}, $\ldots, y_{0n}, x_{0i}, x_{01}, \ldots, x_{0n}$ are vectors, (6.13) is rewritten as

$$x_{1i} = \xi^1 \Delta A \eta^1 x_{01} + \cdots + \xi^n \Delta A \eta^n x_{0n}, \quad (6.14)$$

where the substitutions $\frac{x_{0i}}{\lambda_{0i} - \lambda_{01}} = \eta^1, \ldots, \frac{x_{0i}}{\lambda_{0i} - \lambda_{0n}} = \eta^n$, and $y_{01}^T = \xi^1, \ldots, y_{0n}^T = \xi^n$ are used. Observing that $\xi^1 \Delta A \eta^1, \ldots, \xi^n \Delta A \eta^n$ are scalars, we consider the jth element of the vector x_{1i} to be

$$(x_{1i})_j = \xi^1 \Delta A \eta^1 (x_{01})_j + \cdots + \xi^n \Delta A \eta^n (x_{0n})_j. \tag{6.15}$$

Let us rewrite (6.15) as

$$
\begin{aligned}
(x_{1i})_j &= \left\{ \sum_{k=1}^n \sum_{l=1}^n \left((\xi^1)_k (\Delta A)_{kl} (\eta^1)_l \right) \right\} (x_{01})_j + \\
&\quad \cdots + \left\{ \sum_{k=1}^n \sum_{l=1}^n \left((\xi^n)_k (\Delta A)_{kl} (\eta^n)_l \right) \right\} (x_{0n})_j \\
&= \sum_{p=1}^n \left\{ \sum_{k=1}^n \sum_{l=1}^n \left((\xi^p)_k (\Delta A)_{kl} (\eta^p)_l \right) \right\} (x_{0p})_j \\
&= \sum_{k=1}^n \sum_{l=1}^n \left\{ \sum_{p=1}^n \left((\xi^p)_k (\eta^p)_l \right) (x_{0p})_j \right\} (\Delta A)_{kl}. \tag{6.16}
\end{aligned}
$$

Therefore, since $\sum_{p=1}^n \left((\xi^p)_k (\eta^p)_l \right) (x_{0p})_j$ is a complex number, simply by writing $\alpha_{kl} + \beta_{kl} \mathrm{i} := \sum_{p=1}^n \left((\xi^p)_k (\eta^p)_l \right) (x_{0p})_j$, we have

$$(x_{1i})_j = \sum_{k=1}^n \sum_{l=1}^n (\alpha_{kl} + \beta_{kl} \mathrm{i})(\Delta A)_{kl}. \tag{6.17}$$

Next, in order to find the maximum absolute magnitude of $(x_{1i})_j$, we separate the real part and the imaginary part. Let us first investigate the maximum absolute value of the real part. The real part is $\sum_{k=1}^n \sum_{l=1}^n \alpha_{kl} (\Delta A)_{kl}$, where $\alpha_{kl} \in \Re$ and $(\Delta A)_{kl}$ are scalar intervals. Observe that $\sum_{k=1}^n \sum_{l=1}^n \alpha_{kl} (\Delta A)_{kl}$ is a scalar interval, defined by $[\underline{\delta}, \overline{\delta}]$, where

$$\underline{\delta} = \min_{(\Delta A)_{kl} \in (\Delta A)_{kl}^I} \left\{ \sum_{k=1}^n \sum_{l=1}^n \alpha_{kl} (\Delta A)_{kl} \right\}$$

$$\text{and } \overline{\delta} = \max_{(\Delta A)_{kl} \in (\Delta A)_{kl}^I} \left\{ \sum_{k=1}^n \sum_{l=1}^n \alpha_{kl} (\Delta A)_{kl} \right\},$$

where $(\Delta A)_{kl}^I := \left[\underline{a_{ij}} - a_{ij}^o, \overline{a_{ij}} - a_{ij}^o \right]$. Finally, from the same argument given in Lemma 4.9, we find that $\underline{\delta}$ occurs at a vertex point of $(\Delta A)_{kl}^I$ depending on signs of α_{kl}. In other words, if $\alpha_{kl} \geq 0$, then the minimum of δ occurs at $a_{ij}^o - \underline{a_{ij}}$; else if $\alpha_{kl} < 0$, then the minimum of δ occurs at $\overline{a_{ij}} - a_{ij}^o$. In the same way, $\overline{\delta}$ occurs at a vertex point of $(\Delta A)_{kl}$ depending on signs of α_{kl}. If $\alpha_{kl} \geq 0$, then the maximum of δ occurs at $\overline{a_{ij}} - a_{ij}^o$; else if $\alpha_{kl} <$

0, then the maximum of δ occurs at $a_{ij}^o - a_{ij}$. Next, the same procedure can be repeated for the imaginary part. Thus, the maximum and minimum boundaries of eigenvectors can be checked by investigating the vertex matrices of the interval perturbation matrix.   ∎

Lemma 6.5 and Lemma 6.6 show how the maximum magnitude of the perturbed eigenvalues and eigenvectors can be calculated, respectively. Thus, since the perturbed eigenpairs are calculated by $\lambda_i = \lambda_{0i} + \lambda_{1i}$, $x_i = x_{0i} + x_{1i}$ and $y_i = y_{0i} + y_{1i}$, we have effectively computed the bounds on the interval matrices Λ^I, X^I, and Y^I.

6.3 Markov Parameter Bounds

In Section 6.1, the interval model conversion method was developed and in Section 6.2, an analytical method for finding the maximum magnitudes of the perturbed eigenpairs was suggested. That is, Section 6.2 showed that the interval boundaries of A^k, where $A \in A^I$, can be bounded using the inequality $(A^I)^k \subseteq A_k^I$, which provides the following relationship:

$$\underline{\mathcal{A}^k} \le \underline{A^k} \le A^k \le \overline{A^k} \le \overline{\mathcal{A}^k} \tag{6.18}$$

where $A^k \in (A^I)^k$ and $\mathcal{A}^k \in A_k^I$. Then, the interval boundaries of Markov parameters (e.g., $h_{k+1} = CA^kB$) can be estimated by $\underline{h_{k+1}} = \underline{CA^kB}$; $\overline{h_{k+1}} = \overline{CA^kB}$, where C, B are constant vectors describing the system (6.1) and A^k is a matrix which is lower-bounded and upper-bounded by $\underline{A^k} \le A^k \le \overline{A^k}$ from (6.18). Finally, Lemma 6.5 and Lemma 6.6 of Section 6.2 showed that the analytical solution for estimating the boundaries of the interval matrix $A_k^I := \{M = X\Lambda^kY : \Lambda \in \Lambda^I, X \in X^I, Y \in Y^I\}$ can be obtained using the vertex matrices of interval perturbation matrix ΔA^I.

6.4 Robust ILC Design

Now, let us discuss the synthesis of interval iterative learning control (IILC) systems, assuming we know the boundaries of H.[2] For MC ILC design in the 2-norm topology, we make use of Lemma 4.7. For the ILC update law $U_{k+1} = U_k + \Gamma E_k$, the evolution of the error is given by the following error vector update law: $E_{k+1} = (I - H\Gamma)E_k$, $H \in H^I$ where the singular value of $T = (I - H\Gamma)$ (with interval, $T^I = (I - H^I \otimes \Gamma)$) is calculated as

---

[2] The design we present here is applicable whenever we know the boundaries of H, whether that knowledge was given a priori or was computed using the results given in the previous sections of this chapter.

$$\overline{\sigma}(T) = \rho\left[(T)^T T\right] = \sqrt{\rho\begin{bmatrix} 0 & (T)^T \\ T & 0 \end{bmatrix}} = \sqrt{\rho(\mathbf{P})} \left(\text{def.} \begin{bmatrix} 0 & (T^I)^T \\ T^I & 0 \end{bmatrix} = \mathbf{P}^I\right).$$
$$(6.19)$$

Then, the MC property of the IILC system is checked by analyzing the Schur stability of $\mathbf{P} \in \mathbf{P}^I$ in the 2-norm topology using Lemma 4.7.

For MC conditions in the 1- and ∞-norm topologies, Theorem 4.12 and Theorem 4.13 are used, which is summarized in the following lemma for convenience.

Lemma 6.7. *Given $h_i \in [\underline{h_i}, \overline{h_i}]$, the IILC system is monotonically-convergent if $\|I - H^v \Gamma\|_k < 1$, where k is 1 or ∞; and H^v are vertex Markov matrices associated with the interval Markov matrix H^I.* ∎

For achieving AS, as shown in Theorem 4.8, if Arimoto-like gains are selected such that $|1 - \gamma \underline{h_1}| < 1$ and $|1 - \gamma \overline{h_1}| < 1$, the AS of the IILC system is achieved. To increase robustness, the following scheme is recommended:

$$\gamma = \begin{cases} 1/\overline{h_1} & \text{if } h_1 \geq 0 \text{ for all } h_1 \in h_1^I \\ 1/\underline{h_1} & \text{if } h_1 < 0 \text{ for all } h_1 \in h_1^I \end{cases}.$$
$$(6.20)$$

The reason for Eq. (6.20) is that the AS condition $|1 - \gamma \overline{h_1}|$ is guaranteed with $\gamma = 1/\overline{h_1}$ if $h_1 \geq 0$ for all $h_1 \in h_1^I$ and with $\gamma - 1/\underline{h_1}$ if $h_1 < 0$ for all $h_1 \in h_1^I$. For the MC ILC gain design in the 2-norm topology, we use (6.19). For the learning gain matrix design, the following optimization is suggested based on Lemma 4.7.

Suggestion 6.8. *Let $\mathbf{p}_{ij}^I$ be the ith row and j^{th} column element of $\mathbf{P}^I$. If we define a matrix $\mathbf{M}$ whose elements are given as $m_{ij} = \max\{|\underline{p_{ij}}|, |\overline{p_{ij}}|\}$, then we solve the following optimization problem to design Γ:*

$$\min_{\Gamma} \rho(\mathbf{M}) \quad \text{s.t.} \quad h_k \in [\underline{h_k}, \overline{h_k}].$$
$$(6.21)$$

Remark 6.9. To explain why the matrix $\mathbf{M}$ is used in (6.21), define $\mathbf{s}_{ij}^1$ ($\mathbf{S}^1 = [\mathbf{s}_{ij}^1]$) and $\mathbf{s}_{ij}^2$ ($\mathbf{S}^2 = [\mathbf{s}_{ij}^2]$) as

$$\mathbf{s}_{ij}^1 := \overline{p_{ij}} \text{ if } i = j; \quad \mathbf{s}_{ij}^1 := \max\{|\underline{p_{ij}}|, |\overline{p_{ij}}|\} \text{ if } i \neq j;$$

$$\mathbf{s}_{ij}^2 := \underline{p_{ij}} \text{ if } i = j; \quad \mathbf{s}_{ij}^2 := \min\{-|\underline{p_{ij}}|, -|\overline{p_{ij}}|\} \text{ if } i \neq j.$$

Then, since the matrix $\mathbf{P}^I$ is symmetric, $\mathbf{S}^1 = -\mathbf{S}^2$. Hence, using the fact that $\rho(\mathbf{S}^1) = \rho(\mathbf{S}^2)$ and the diagonal terms of $\mathbf{P}^I$ are all zero, we only need to check the spectral radius of the matrix composed of the off-diagonal terms of $\mathbf{S}^1$.

If Lemma 6.7 is used then a non-constrained optimization scheme is suggested:

Suggestion 6.10. *If k is 1 or ∞, the following optimization is straightforward:*

$$\min_{\Gamma} \| I - H^v \Gamma \|_k. \tag{6.22}$$

The optimization in Suggestion 6.10 is to minimize $\| I - H^v \Gamma \|_k$ using Γ, where Γ is a band-fixed learning gain matrix. There is a trade-off. For a small band size, it is possible that there might not exist an optimization solution such that $\| I - H^v \Gamma \|_k < 1$. In this case, the band size should be increased until the optimization algorithm finds a Γ such that $\| I - H^v \Gamma \|_k < 1$. However, as the band size increases, more causal and noncausal learning gains are required. This is practically undesirable because we need to store more data into memory for the current control update. The optimization in Suggestion 6.8 is a nonlinear constrained minimization problem and the optimization in Suggestion 6.10 is a nonlinear non-constrained minimization problem. These problems can be easily solved using the MATLAB® Optimization Toolbox. Depending on the IILC system, the optimization scheme suggested above may not find the optimization solution even with fully populated learning gain matrix. In this case, the following control update law could be used: $U_{k+1} = Q(U_k + \Gamma E_k)$, where Q is a time-invariant diagonal matrix.[3] Then, since the error vector is updated by the following formula: $E_{k+1} = Q(I - H\Gamma)E_k + (I - Q)Y_d$, it is easy to make $\| Q(I - H\Gamma) \| < 1$ by choosing Q and Γ. However, the remaining term $(I - Q)Y_d$ will result in a nonzero steady-state error. This is a trade-off.

6.5 Simulation Illustrations

For the verification of the interval model conversion idea developed in this chapter, we present the results of a simulation test. Let us use the following simple discrete servo system model, whose nominal plant was identified from the Quanser SRV02 system, as discussed in the previous chapter's simulation example: [4]

$$x_1(k+1) = a_{11}x_1(k) + a_{12}x_2(k) + 2u(k) \tag{6.23}$$

$$x_2(k+1) = a_{21}x_1(k) + a_{22}x_2(k) + 0.5u(k) \tag{6.24}$$

$$y(k) = x_1(k), \tag{6.25}$$

where the interval plant parameters are bounded as $-0.74 \leq a_{11} \leq -0.66$, $-0.53 \leq a_{12} \leq -0.47$, $0.95 \leq a_{21} \leq 1.05$, and $0.19 \leq a_{22} \leq 0.21$. Using the results of the foregoing sections, a simulation test is performed with the following reference sinusoidal signal: $Y_d = \sin(8.0j/n)$, where $n = 20$ and $j = 1, \dots, n$. The band size is fixed to be 3. Figure 6.1 shows the calculated interval boundaries of the Markov parameters using the computations given in Sections 6.1 and 6.2. The circle-marked line represents the maximum/minimum boundaries

---

[3] The use of the so-called Q-filter is well-known in the ILC literature. See [332], for example.

[4] http://www.quanser.com/english/html/challenges/fs_chall_rotary_flash.htm.

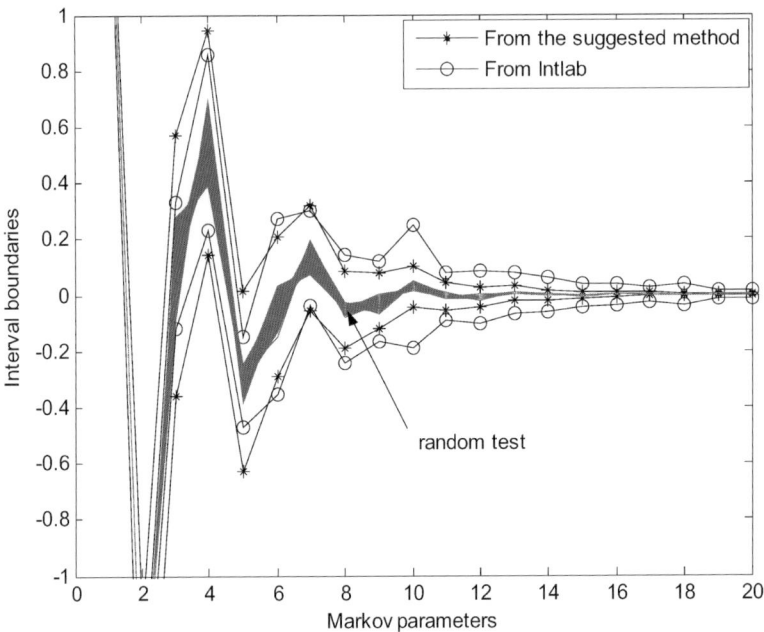

Fig. 6.1. Calculated Markov parameter interval uncertain boundaries

of the Markov parameters calculated from the Intlab software [179]; and the
∗-marked line represents the Markov parameter boundaries calculated from
the method suggested in this section. For verification of the suggested method,
a Monte-Carlo type random test was also performed as shown in the figure
(identified by "random test" in the figure). Observe that the suggested method
gives reliable bounds for the interval ranges of the Markov parameters. We see
that the suggested method gives less conservative bounds than Intlab after h_6;
but, from h_1 to h_5, Intlab is slightly less conservative. However, we note that
we have found in other experiments that in the case of a marginally stable
system, the suggested method is less conservative than Intlab for all Markov
parameters. So the suggested method is particularly suitable for MC ILC design. We should further stress that the technique proposed in this section gives
an analytical computation of the bounds for all h_k, unlike the bounds found
from Intlab. To see the convergence of the ILC process for a single plant in the
interval system, refer to Figure 6.2. The dot-marked line (called the "interval
matrix method") is the maximum l_2 norm error of the ILC process, whose
learning gain matrix was designed by (6.21); and the circle-marked line (called
the "norm-based method") is the maximum l_1 norm error when the learning
gain matrix was designed by (6.22). In the case of the interval matrix method,
there is a steady-state error, because we fixed $Q = 0.9$ to guarantee the MC

as commented in Section 6.4 (when $Q = 1$, the optimization did not find the optimal solution such that the norm is less than 1).

6.6 A Different Approach for Interval Model Conversion

In the preceding section, for the calculation of the power of interval matrix, the eigen-decomposition method was used. Although this method is effective and computational effort is relatively small, the result could be conservative, depending on the system. In Appendix D, a new method for computing the power of an interval matrix is developed, using the sensitivity transfer from the nominal to the perturbed power of the interval matrix. This method provides accurate boundaries of the power of an interval matrix, but requires a greater amount of computation. Therefore, there is a trade-off between using the eigen-decomposition method presented above and using the sensitivity transfer method explained in Appendix D. In this monograph, because the power of interval matrix is a fundamental research topic under robust control, we provide this new method in a separate appendix. It is straightforward to apply the result given in Appendix D to the IILC system and we do not present the details here.

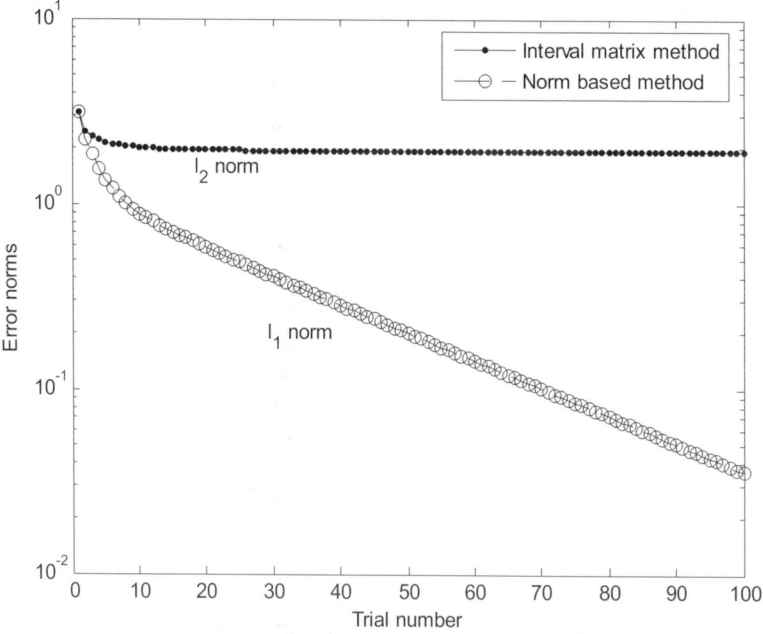

Fig. 6.2. ILC convergence test results from both the interval matrix method and the norm-based method

6.7 Chapter Summary

In this chapter, a robust iterative learning controller was designed for the case where the interval boundaries of the system Markov matrix are assumed to be known. A method was given to convert interval uncertainty in the plant's state-space A-matrix into boundaries for the associated system Markov matrix. The method only requires computations using the vertex matrices of the interval system. Optimization schemes were also suggested based on an interval matrix stability analysis method and a norm-based method. In the case of the norm-based method, the ILC learning gain matrix guaranteed the MC of the uncertain ILC process with zero steady-state error. However, the interval matrix method only guaranteed MC with nonzero steady error. From these results, it is concluded that the norm-based method is less conservative than the interval matrix method. However, the norm-based method requires much more computational time than the interval matrix method.

Part III

Iteration-domain Robustness

7

Robust Iterative Learning Control:
H_∞ Approach

In Chapter 4, Chapter 5, and Chapter 6, robust monotonic convergence (MC) of first-order ILC (FOILC) was studied for the case when the time-domain plant is subject to parametric interval uncertainty. In this chapter and the next we consider ILC design for the case when the plant is subject to iteration-domain uncertainty.

We begin in this chapter by presenting an H_∞-based design technique for synthesis of higher-order iterative learning controllers (HOILC) for when there are iteration-domain input/output disturbances and plant model uncertainty. By formulating the HOILC problem as a high-dimensional MIMO discrete-time system as in [301], it is shown how the problem of input/output disturbances and plant model uncertainty can be cast into a standard H_∞ framework. An algebraic approach for solving the problem in this framework is used, based on [142], resulting in a sub-optimal controller that can achieve both stability and robust performance. The key observation is that H_∞ synthesis can be used for HOILC design to achieve reliable performance in the presence of iteration-varying external disturbances and model uncertainty. As commented in Section 3.3, however, it is difficult to guarantee the monotonic convergence of HOILC system. Hence this chapter focuses on asymptotic stability conditions.

7.1 Introduction

As we have said, iterative learning control (ILC) is a technique that attempts to refine the performance of systems that repeat their operation over and over from the same initial conditions. Hence ILC is fundamentally a two-dimensional process, with evolution along both a finite time axis (i.e., finite horizon; denoted by t) and an infinite iteration axis (i.e., infinite horizon) (denoted by k) [301]. As shown in Chapter 2, many ILC publications have considered robustness issues, including H_∞-based ILC design from the time-axis perspective [102, 103, 321], stochastic ILC design [395], and disturbance rejection [236, 58, 531, 268]. However, for the most part, existing works have focused on ILC design for performance improvement, with the assumption that the plant is iteration-invariant and with external disturbances treated along the time axis. That is, the primary focus has been on robustness defined and modeled along the time axis. To date, iteration-axis robustness has not been treated in a systematic way (one exception to this, in a somewhat different framework, is the work on multi-pass systems [387, 126]). In this chapter, a new framework is suggested for robust ILC design, assuming both time-varying model uncertainty and iteration-varying external disturbances. Using the super-vector approach we can easily incorporate iteration-varying disturbances and the system can be analyzed using discrete (iteration-axis) frequency-domain techniques.

7.2 Problem Formulation

We will use the basic ILC dynamics (1.22–1.27) given in Chapter 1. As in [301], we introduce a delay operator along the iteration axis, w, with the property that $w^{-1}u_k(t) = u_{k-1}(t)$ (note, this delay operator was already defined in Section 3.1). We call this the "w-transform." It operates from trial-to-trial, with t fixed, as opposed to the standard z-transform operator that operates from time step-to-time step, with k fixed. Thus, $Y_k = HU_k$ can be written as $Y_k(w) = HU_k(w)$.

Now consider the general form of a (higher-order) ILC algorithm:

$$U_{k+1} = -\bar{D}_n U_k - \bar{D}_{n-1} U_{k-1} - \cdots - \bar{D}_0 U_{k-n}$$
$$+ N_n E_k + \cdots + N_1 E_{k-n+1} + N_0 E_{k-n}. \tag{7.1}$$

Here the "next input" is computed as a filtered sum of all the "past inputs" and "past errors." Taking the w-transform of both sides of this equation and combining terms gives

$$\bar{D}_c(w)U(w) = N_c(w)E(w), \tag{7.2}$$

where

$$\bar{D}_c(w) = w^{n+1} + \bar{D}_n w^n + \cdots + \bar{D}_1 w + \bar{D}_0 \tag{7.3}$$

$$N_c(w) = N_n w^n + N_{n-1} w^{n-1} + \cdots + N_1 w + N_0, \tag{7.4}$$

which can also be written in a matrix fraction as $U(w) = \bar{C}(w)E(w)$ where

$$\bar{C}(w) = \bar{D}_c^{-1}(w)N_c(w). \tag{7.5}$$

Note that a common special case of (7.1) uses the general (higher-order) ILC update law

$$U_{k+1} = (I - D_{n-1})U_k + (D_{n-1} - D_{n-2})U_{k-1}$$
$$+ \cdots + (D_1 - D_0)U_{k-n+1} + D_0 U_{k-n}$$
$$+ N_n E_k + \cdots + N_1 E_{k-n+1} + N_0 E_{k-n}. \qquad (7.6)$$

Taking the w-transform of the update law (7.6) yields

$$(w - 1)D_c(w)U(w) = N_c(w)E(w), \qquad (7.7)$$

where

$$D_c(w) = D_n w^n + D_{n-1} w^{n-1} + \cdots + D_1 w + D_0 \qquad (7.8)$$
$$N_c(w) = N_n w^n + N_{n-1} w^{n-1} + \cdots + N_1 w + N_0, \qquad (7.9)$$

which can also be written in a matrix fraction as

$$U(w) = \frac{I}{(w-1)} C(w)E(w), \qquad (7.10)$$

where

$$C(w) = D_c^{-1}(w)N_c(w). \qquad (7.11)$$

Figure 7.1 depicts the set of equations we have just developed based on (7.1), for a general ILC update law represented by $\bar{C}(w)$, or (7.6), for an ILC update law represented by $(w - 1)^{-1}C(w)$. From this figure it is clear that the repetition-domain closed-loop dynamics become either:

$$\bar{G}_{cl}(w) = H[\bar{D}_c(w) + N_c(w)H]^{-1}N_c(w) \qquad (7.12)$$

for (7.1) or, for (7.6),

$$G_{cl}(w) = H[(w - 1)D_c(w) + N_c(w)H]^{-1}N_c(w). \qquad (7.13)$$

For the latter case, because we now have an integrator in the feedback loop (a discrete integrator, in the repetition domain), applying the final value theorem to G_{cl} shows that $E_k \to 0$ as long as the ILC algorithm converges (i.e., as long as G_{cl} is stable).

7.2.1 A Generalized Framework

The development given above can be extended in a number of ways, by including:

1. Iteration-varying reference signals $Y_d(w)$.
2. Iteration-varying noise signals $N(w)$.
3. Iteration-varying disturbance signals $D(w)$.
4. Iteration-varying nominal plant models $H(w)$.
5. Iteration-varying plant model uncertainty $\Delta H(w)$.

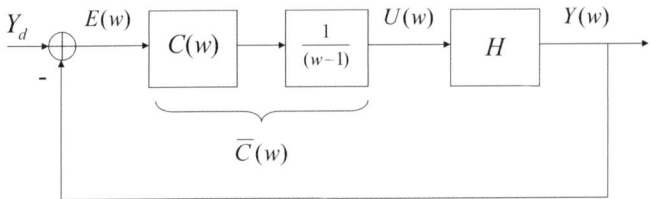

Fig. 7.1. Standard ILC setup in the super-vector framework

6. Separation of the control action into current-cycle feedback C_{CITE} and ILC update C_{ILC} feedback.

Figure 7.2 depicts the complete picture. Note that in this figure we have assumed that integrating action (in the iteration domain) is used in the control law. Further, the diagram shows the current-cycle feedback separated from the ILC update. Both of these effects can be absorbed into a single controller denoted by $\bar{C}(w)$. In particular, note that the use of current cycle feedback can be incorporated into our algorithm by simply adding a term $N_{k+1}E_{k+1}$ to both (7.1) and (7.6). This simply means that with respect to our iteration-domain feedback system, the controller now has relative degree zero rather than the relative degree one controller that results when only previous cycle feedback is used.

The important point we want to stress here is that the framework depicted in Figure 7.2 emphasizes the role of iteration-variant effects in the ILC problem formulation. In the remainder of this chapter we consider the special case of Figure 7.2 when there is no current-cycle feedback, the signals $D(w)$ and $N(w)$ have known l_2-norm bounds (in the iteration domain), and the model uncertainty ΔH has a known H_∞ system norm bound.

7.2.2 Iteration-domain H_∞ Problem Formulation

Figure 7.1 depicts the general higher-order ILC problem as a MIMO control problem with the plant H. This figure highlights the fact that the ILC process (i) is inherently a relative degree one process; and (ii) should have an integrator in order to converge to zero steady-state error. However, it can also be noted that the controller $C(w)$ has relative degree zero. This makes it convenient to consider the reformulation shown in Figure 7.3. Here:

1. The integrator has been grouped with the plant, so that we now define a new plant

$$H_p(w) = (w - 1)^{-1}H. \tag{7.14}$$

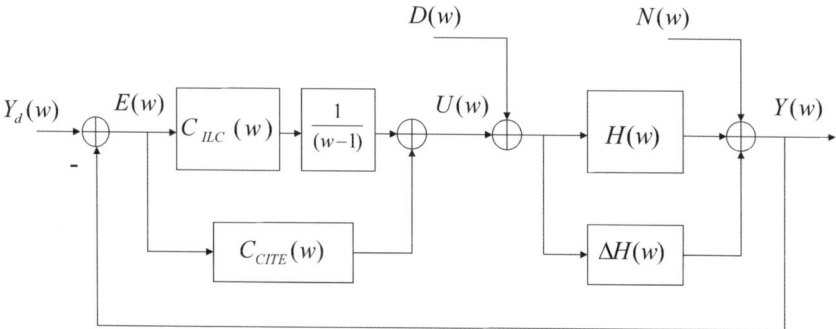

Fig. 7.2. More general ILC framework

2. We suppose that H is subject to a perturbation such as $H = H_0 + \Delta H(w)$ where $\Delta H(w)$ represents iteration-varying uncertainty in the plant model.

3. The plant is disturbed by a plant input disturbance d_I and a plant output disturbance d_o.

These disturbances and plant perturbation models lead to a standard H_∞ problem. However, before proceeding, we must first define our signal spaces properly. To this end we introduce the following definition:

Definition 7.1. *In super-vector ILC, the l_2-norm should be defined along the iteration axis (i.e., on the w-domain). To distinguish this concept from the discrete-time domain, the iteration-domain l_2-norm is written as $\| \cdot \|_{2^w}$ and denoted l_{2^w}. Thus, rather than writing $\|E_k\|_2$, we use the notation $\|E_k\|_{2^w}$. In the same way, the iteration-domain H_∞ system norm is written as $\| \cdot \|_{\infty^w}$*
∎

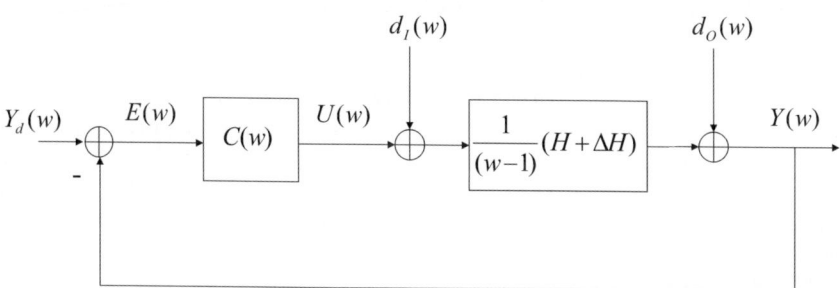

Fig. 7.3. ILC H_∞ problem framework

The design problem for the uncertain ILC system depicted in Figure 7.3 can now be formulated as follows:

Problem: Given $H_p(w) = (w-1)^{-1}H$ and $Y_d(w)$, find $C(w)$ in Figure 7.3 such that $\|E_k\|_{2^w}$ is minimum from the l_{2^w}-bounded disturbances d_I and d_o and the closed-loop system is stable over all $H = H_0 + \Delta H(w)$, with $\|\Delta H(w)\|_{\infty^w} < \epsilon_H$.

A number of remarks are in order:

Remark 7.2. Notice that the problem setup we have defined indicates that the higher-order ILC system can be synthesized in the H_∞ framework. That is, the minimization problem of $\|E(w)\|_{2^w}$ is translated into the minimization problem

$$\|T_{EW}\|_{\infty^w} = \sup_{W_k \in l_{2^w}} \frac{\|E_k\|_{2^w}}{\|W_k\|_{2^w}}, \tag{7.15}$$

such that

$$\|T_{EW}\|_{\infty^w} < \gamma, \tag{7.16}$$

where[1]

$$W_k = [d_I, d_o]^T \tag{7.17}$$

$$\|E_k\|_{2^w} = \sum_{k=0}^{k=\infty} E_k^T E_k, \tag{7.18}$$

$$\|W_k\|_{2^w} = \sum_{k=0}^{k=\infty} W_k^T W_k. \tag{7.19}$$

When γ is not fixed, this is an optimal H_∞ ILC problem and when γ is fixed this a sub-optimal H_∞ ILC problem. In the remainder of the chapter we design the ILC controller $C(w)$ with fixed $\gamma = 1$.

Remark 7.3. The minimization of T_{EW} here means the reduction of the H_∞ norm of the transfer matrices from W_k to E_k. In other words, the reference input Y_d is not counted in this performance problem. So, minimization of T_{EW} does not guarantee the minimization of $E_k = Y_d - Y_k$. Instead, the sensitivities of d_I and d_o to E_k are reduced by minimization of $\|T_{EW}\|_{\infty^w}$. The minimization of $E_k = Y_d - Y_k$ is ensured by the presence of the integrator in the loop gain and by adequate solution of the robust stability problem.

Remark 7.4. $H_p(w)$ has what is called a structured perturbation, because we have

---

[1] Note that d_I and d_o are iteration-varying signals and thus so is $W = [d_I, d_o]^T$. A more formally correct notation for d_I and d_o would be $(d_I)_k$ and $(d_o)_k$. However, we will drop the dependence on k for convenience when the meaning is clear from the context.

$$H_p(w) = (w-1)^{-1}H = \frac{I}{(w-1)}\left(H_0 + \Delta H(w)\right),\qquad(7.20)$$

not

$$H_p = H_0 + \Delta H(w).$$

That is, there is no modeling uncertainty associated with the integrator, as the integrator is actually part of the controller, not the plant.

Remark 7.5. Figure 7.3, with (7.16), defines a classic H_∞ robust control problem. From this framework we can formulate and solve any of the associated robust stability or robust performance problems with an H_∞ optimality criteria. Such solutions can be obtained using any number of standard techniques. In this chapter, we apply an algebraic approach from the literature [142], which has the advantage of providing an analytical solution. However, we also compare the resulting solution with that obtained numerically using the standard MATLAB® *dhinf* command.

7.3 Algebraic H_∞ Approach to Iterative Learning Control

In this section, we consider the robust design problem for external disturbances and for model uncertainty separately.

7.3.1 Iteration-varying Disturbances

Figure 7.4 shows the block diagram of the general H_∞ problem cast into the form used to give the so-called algebraic solution, presented in [142], with the plant written in state-space form as

$$x_{k+1} = Ax_k + B_1 w_k + B_2 u_k \qquad(7.21)$$

$$z_k = C_1 x_k + D_{11} w_k + D_{12} u_k \qquad(7.22)$$

$$y_k = C_2 x_k + D_{21} w_k + D_{22} u_k, \qquad(7.23)$$

where $z_k = [z_k^1, z_k^2]^T$ is the performance output, y_k is the observation output to be used in output feedback control, and $w_k = [d_I, d_o]^T$ are the exogenous inputs (disturbances in plant input and plant output). Generally, it is convenient to assume that $D_{11} = 0$ (or, D_{11} should be very small to hold the existence condition) and $D_{22} = 0$. In Figure 7.4, W_1 and W_2 are penalty weighting matrices, and W_i and W_o are disturbance-generating functions. In the ILC problem, W_1 and W_2 are identity matrices, and $W_i = W_o = \alpha I$.

For a solution K to exist in Figure 7.4, we need the following assumptions:
A1: (A, B_2) is stabilizable and (C_2, A) is detectable.
A2: D_{12} is full column rank and D_{21} is full row rank.
The controller existence condition such that $\|T_{zw}\|_\infty < 1$ while stabilizing the system is given in many references, see [269], for example, and a solution can

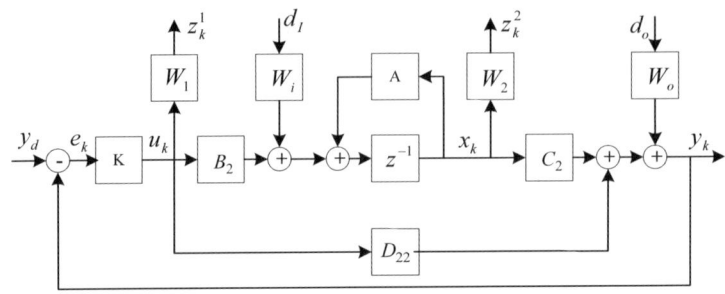

Fig. 7.4. Typical discrete H_∞ diagram

be given directly from MATLAB® using the *dinf* command. Here, however, we will follow the approach in [142] that gives an analytical solution.

Now, we show that the HOILC problem can be cast into the standard H_∞ framework of Figure 7.4. From Figure 7.3, with $\Delta H(w) = 0$, the HOILC system can be written as

$$U_{k+1} = U_k + V_k + d_I \tag{7.24}$$
$$Y = HU_k + d_o, \tag{7.25}$$

where we have defined $V_k(w) = C(w)E(w)$.

To proceed, we will reformulate (7.24)–(7.25) into a state-space form corresponding to (7.21), (7.22), and (7.23), from which we can redraw Figure 7.3 into the form of Figure 7.4. The outcome of this will be Figure 7.5. When done, in Figure 7.5, the performance outputs will be selected as $Z_k = [z_k^1, z_k^2]^T = [E_k, D_{12}V_k]^T$, where $V_k = C(w)E_k$, so that the input plant sensitivity matrix (from d_I to E_k) becomes

$$S_I = H_p(w)(I + C(w)H_p(w))^{-1} \tag{7.26}$$

and the output sensitivity matrix becomes

$$S_o = (I + C(w)H_p(w))^{-1}. \tag{7.27}$$

In this performance penalty, $D_{12}V_k$ is used to satisfy the full column rank requirement and also to minimize the control effort due to the external disturbance. Therefore, the H_∞ performance problem is to minimize two sensitivity matrices (i.e., S_I and S_o) and to stabilize the ILC system (robust stability problem). These two goals define the robust ILC performance problem.

Continuing, the following state-space form of the ILC equations can be derived:

$$U_{k+1} = IU_k + B_1W_k + IV_k \tag{7.28}$$
$$Z_k = [H, 0]^T U_k + D_{11}W_k + D_{12}V_k \tag{7.29}$$
$$Y_k = HU_k + D_{21}W_k, \tag{7.30}$$

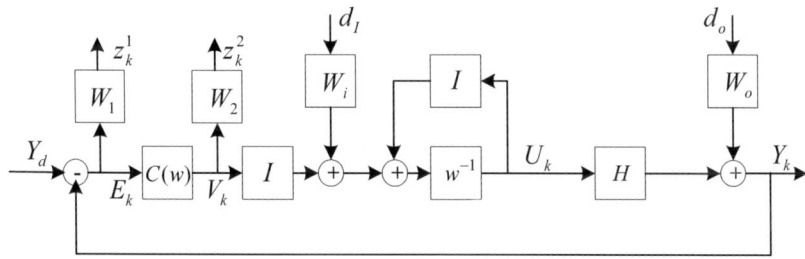

Fig. 7.5. ILC H_∞ diagram with plant input and output disturbances

where H is the Markov matrix, $W_k = [d_I, d_o]^T$, $B_1 = [\alpha I \ 0]$, $V_k = C(w)E_k$, $D_{11} = \begin{bmatrix} 0 & \alpha I \\ 0 & 0 \end{bmatrix}$, and $D_{21} = [0, \ \alpha I]^T$ (α is used to limit the disturbance intensity). Notice that there is a one-to-one mapping between equations (7.21–7.23) and (7.28–7.30) and also between Figure 7.4 and Figure 7.5. Hence, we can apply the algebraic H_∞ solution given in [142] to the higher-order SVILC description, because $(A, B_2)=(I, I)$ is stabilizable and $(C_2, A)=(H, I)$ is detectable with any H. As in the standard H_∞ problem setup, in ILC the following state-space formula can be suggested:

$$A = I; \ B_1 = [\alpha I \ 0]; \ B_2 = I; \ C_1 = [H, \ 0]^T; \ C_2 = H;$$

$$D_{11} = \begin{bmatrix} 0 & \alpha I \\ 0 & 0 \end{bmatrix}; \ D_{12} = [0, \ I]^T; \ D_{21} = [0, \ \alpha I]; \ D_{22} = 0.$$

With this setup, the H_∞-based design of the super-vector ILC controller becomes a typical H_∞ synthesis problem.

For an analytical solution, the following definitions are provided. Let

$$\Theta_1 := \begin{bmatrix} I - \alpha^2 \overline{X} & 0 \\ 0 & I - \alpha^2 I \end{bmatrix}$$

$$\widetilde{B}_1 := [\alpha \overline{X}, 0]^T; \ \widetilde{B}_2 := \overline{X} + H; \ \widetilde{B}_3 := [\alpha \overline{X}, \alpha I]^T$$

$$\Theta_2 := I + \overline{X} + \alpha^2 \overline{X}(I - \alpha^2 \overline{X})^{-1} \overline{X} + \frac{\alpha^2}{1 - \alpha^2} I$$

$$\Theta_3 := H + \overline{X} + \alpha^2 \overline{X}(I - \alpha^2 X)^{-1} \overline{X}$$

$$\overline{A} := I + \alpha^2 (I - \alpha^2 \overline{X})^{-1} \overline{X}$$

$$\overline{B}_1 := [\alpha(I - \alpha^2 \overline{X})^{-\frac{1}{2}}, 0]$$

$$\overline{B}_2 := I + \alpha^2 (I - \alpha^2 \overline{X})^{-1} \overline{X}$$

$$\overline{C}_1 := \Theta_2^{-\frac{1}{2}} \Theta_3; \ \overline{C}_2 := H; \ \overline{D}_{21} := \frac{\alpha}{\sqrt{1 - \alpha^2}} I$$

$$\overline{D}_{22} := \frac{\alpha^2}{1 - \alpha^2} I$$

$$\Theta_4 := I - \overline{C}_1 \overline{Z} \overline{C}_1^T$$
$$E_1 := \overline{C}_1 \overline{Z} \overline{C}_1^T$$
$$E_2 := \overline{C}_2 \overline{Z} \overline{A}^T + \overline{D}_{21} \overline{B}_1^T$$
$$E_3 := \overline{C}_1 \overline{Z} \overline{C}_2^T$$
$$\Theta_5 := \overline{D}_{21} \overline{D}_{21}^T + \overline{C}_2 \overline{Z} \overline{C}_2^T + E_3^T \Theta_4^{-1} E_3$$
$$\Theta_6 := E_2 + E_3^T \Theta_4^{-1} E_1.$$

With these definitions, the following theorem is suggested.

Theorem 7.6. *For the HOILC system given by (7.28), (7.29), (7.30), if there exists $\overline{X} > 0$ and $F = -\Theta_2^{-1}\Theta_3$ such that $\Theta_1 > 0$ and $I + \alpha^2 \overline{X}(I - \alpha^2 \overline{X})^{-1}\overline{X} - \Theta_3^T \Theta_2^{-1} \Theta_3 < 0$, and also if there exists $\overline{Z} > 0$ such that $\Theta_4 > 0$ and $\overline{AZA}^T - \overline{Z} + \overline{B}_1 \overline{B}_1^T + E_1^T \Theta_4^{-1} E_1 - \Theta_6^T \Theta_5^{-1} \Theta_6 < 0$, then the observer gain matrix $L := -\Theta_6^T \Theta_5^{-1}$ and the controller given as*

$$\xi_{k+1} = A_c \xi_k + B_c Y_k \tag{7.31}$$
$$V_k = C_c \xi_k \tag{7.32}$$

where $A_c := \overline{A} + (\overline{B}_2 + L\overline{D}_{22})F + L\overline{C}_2$, $B_c = L$, and $C_c = F$, stabilize the system (7.28), (7.29), (7.30) and guarantee $\|T_{ZW}\|_{\infty^w} < 1$. ∎

Proof. The proof is immediate following several algebraic simplifications after substituting

$$A = I; \ B_1 = [\alpha I \ 0]; \ B_2 = I; \ C_1 = [H, \ 0]^T; \ C_2 = H;$$

$$D_{11} = \begin{bmatrix} 0 & \alpha I \\ 0 & 0 \end{bmatrix}; \ D_{12} = [0, \ I]^T; \ D_{21} = [0, \ \alpha I]; \ D_{22} = 0$$

into the system given in Theorem 3.3 of [142]. ∎

Remark 7.7. In the controller defined in Theorem 7.6, the dimensions of A_c, B_c, and C_c are the same as that of the Markov matrix H. This implies that the controller $C(w)$ will be of order N in iteration.

Remark 7.8. To solve for the controller matrices A_c, B_c and C_c, in Theorem 7.6, two discrete Riccati-type equations, (7.33) and (7.34), have to be solved to find the positive definite matrices $\overline{X}$ and $\overline{Z}$. From Remark 3.2 of [142] we can use a recursive solution searching method to solve these Riccati equations. The following equations can be used for this purpose:

$$(\overline{X})_{i+1} = (\overline{X})_i + H^T H + \alpha^2 \overline{X}(I - \alpha^2 \overline{X})^{-1}\overline{X} + \varepsilon I, \tag{7.33}$$

$$(\overline{Z})_{i+1} = (\overline{Z})_i + \alpha^2 I + E_1 \Theta_1^{-1} E_1 - \Theta_6^T \Theta_5^{-1} \Theta_6 + \varepsilon I, \tag{7.34}$$

where ε is fixed as 10^{-5}. These equations iterate starting with $(\overline{X})_0 = (\overline{Z})_0 = 0$, under the condition that all conditions of Theorem 7.6 are satisfied. The final outcome of the iteration is chosen as the solution of Riccati the equations.

7.3.2 Model Uncertainty

Next, we extend the results of the previous section to include model uncertainty ΔH in the system (7.28), (7.29), (7.30). Figure 7.6 shows the ILC block diagram with model uncertainty and with the performance output given as $Z_k = D_{12}V_k + C_1U_k$. In this case, the input plant sensitivity matrix (from d_I to Z_k) is defined as

$$S_I = C(w)H_p(w)(I + C(w)H_p(w))^{-1} + (wI - I)^{-1}(I + C(w)H_p(w))^{-1} \tag{7.35}$$

and the output plant sensitivity matrix (from d_o to Z_k) is defined as

$$S_0 = C(w)(I + C(w)H_p(w))^{-1} + C(w)((wI - I)(I + H_p(w)C(w)))^{-1}. \tag{7.36}$$

Then, the purpose of H_∞ synthesis is to minimize these sensitivity matrices and to robustly stabilize the system with uncertainty ΔH. From Figure 7.6, the ILC system can be expressed in the state-space form as

$$U_{k+1} = IU_k + B_1W_k + IV_k \tag{7.37}$$

$$Z_k = C_1U_k + D_{11}W_k + D_{12}V_k \tag{7.38}$$

$$Y_k = (H + \Delta H)U_k + D_{21}W_k + D_{22}V_k, \tag{7.39}$$

where $Z_k = [z_k^1, z_k^2]^T$, $W_k = [d_I, d_o]^T$, $A = I$, $B_1 = [\alpha I, 0]$, $B_2 = I$, $C_1 = [0, I]^T$, $D_{11} = 0_{2n \times 2n}$, $D_{12} = [I, 0]^T$, $C_2 = H + \Delta H$, $D_{21} = [0, \alpha I]$, and $D_{22} = 0$. In this case (I, I) is stabilizable, (H, I) is detectable, D_{12} is full column rank, and D_{21} is full row rank. Thus all the basic assumptions for the existence of an H_∞ solution are satisfied.

Remark 7.9. In Figure 7.5, if model uncertainty is included in H, then the system is changed to:

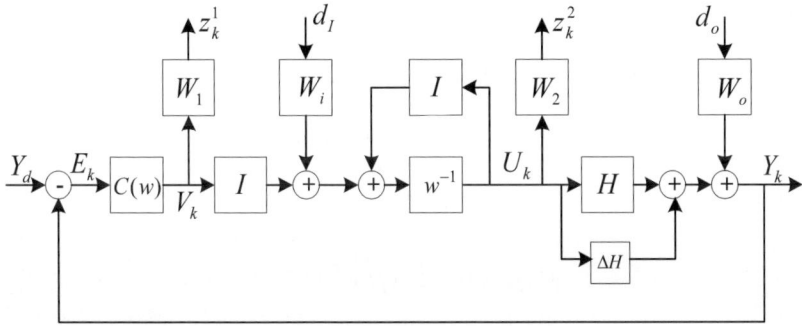

Fig. 7.6. ILC H_∞ framework with model uncertainty

$$U_{k+1} = IU_k + B_1W_k + IV_k \qquad (7.40)$$

$$Z_k = (H + \Delta H)U_k + D_{11}W_k + D_{12}V_k \qquad (7.41)$$

$$Y_k = (H + \Delta H)U_k + D_{21}W_k. \qquad (7.42)$$

Thus, in this case, the performance output Z_k depends on ΔH. But this makes it difficult to standardize the ILC in a typical H_∞ framework. To overcome this drawback, we use Figure 7.6.

To solve the H_∞ robust performance problem algebraically with model uncertainty, an existing result can be used, which is summarized in what follows. From [104, 548], with $D_{22} = 0$, the following uncertain system is given:

$$x_{k+1} = (A + \Delta A)x_k + B_1w_k + (B_2 + \Delta B_2)u_k \qquad (7.43)$$

$$z_k = C_1x_k + D_{11}w_k + D_{12}u_k \qquad (7.44)$$

$$y_k = (C_2 + \Delta C_2)x_k + D_{21}w_k. \qquad (7.45)$$

After making an equivalent auxiliary system given in equations (7.46), (7.47), and (7.48) below, the robust stability analysis can be performed as done above for the system given by equations (7.28)–(7.30). Specifically, let the uncertain system have the following form:

$$\begin{bmatrix} \Delta A & \Delta B_2 \\ \Delta C_2 & \Delta D_{22} \end{bmatrix} = \begin{bmatrix} M_1 \\ M_2 \end{bmatrix} F \begin{bmatrix} N_1 & N_2 \end{bmatrix}$$

where M_1, M_2, N_1, and N_2 are constant matrices, and $FF^T \leq I$. For a parameter $\epsilon > 0$, by introducing the auxiliary system

$$x_{k+1}^a = Ax_k^a + H_1w_k^a + B_2u_k^a \qquad (7.46)$$

$$z_k^a = E_1x_k^a + D_{11}w_k^a + E_2u_k^a \qquad (7.47)$$

$$y_k^a = C_2x_k^a + H_2w_k^a + D_{22}u_k^a, \qquad (7.48)$$

where $H_1 = [\sqrt{\epsilon}M_1, B_1]$, $H_2 = [\sqrt{\epsilon}M_2, D_{21}]$,

$$E_1 = \begin{bmatrix} \frac{N_1}{\sqrt{\epsilon}} \\ C_1 \end{bmatrix}, \text{ and } E_2 = \begin{bmatrix} \frac{N_2}{\sqrt{\epsilon}} \\ D_{12} \end{bmatrix},$$

then the original uncertain system is stabilizable with $\|G_{zw}\| < 1$ if and only if the auxiliary system is stabilizable with $\|G_{z^aw^a}\|_\infty^w < 1$.

Applying this result to our problem, from Figure 7.6 define

$$\begin{bmatrix} 0 & 0 \\ \Delta H & 0 \end{bmatrix} = \begin{bmatrix} 0 \\ \Delta H \end{bmatrix} I \begin{bmatrix} I & 0 \end{bmatrix},$$

where $M_1 = 0$, $M_2 = \Delta H$, $F = I$, $N_1 = I$, and $N_2 = 0$. Then typical H_∞ synthesis can be performed with the augmented ILC system given as

$$x_{k+1}^a = Ax_k^a + B_1 w_k^a + B_2 u_k^a \qquad (7.49)$$

$$z_k^a = C_1 x_k^a + D_{11} w_k^a + D_{12} u_k^a \qquad (7.50)$$

$$y_k^a = C_2 x_k^a + D_{21} w_k^a + D_{22} u_k^a, \qquad (7.51)$$

where $A = I$, $B_1 = [0, \alpha I, 0]$, $B_2 = I$, $C_1 = [\frac{I}{\sqrt{\epsilon}}, 0, I]^T$, $D_{11} = 0_{3n \times 3n}$, $D_{12} = [0, I, 0]^T$, $C_2 = H$, $D_{21} = [\sqrt{\epsilon}\Delta H, 0, \alpha I]$, and $D_{22} = 0$. As done in Theorem 7.6 of this section, Theorem 3.3 of [142] can be modified. This process is quite messy, but straightforward, and is left as an exercise for the (diligent!) reader.

7.4 Simulation Illustrations

In this section we illustrate the performance of the robust ILC controller designed using the H_∞ framework presented in this chapter. Consider the following discrete-time system:

$$x_{k+1} = \begin{bmatrix} 0.25 & 0.6 \\ 0.6 & 0 \end{bmatrix} x_k + \begin{bmatrix} 1.0 \\ 0.0 \end{bmatrix} u_k \qquad (7.52)$$

$$y_k = [1.0 \ -1.3]x_k, \qquad (7.53)$$

which has nominal time-domain eigenvalues at -0.5 and 0.75. In the time axis, 50 discrete samples are used and in the iteration axis, 100 iteration tests are performed. The system starts with the same initial conditions on each iteration. For this system, after calculating the Markov matrix H, the maximum model uncertainty is assumed to be 10 percent of the nominal H (note, for convenience we do not pick up plants with uncertainty $\Delta H(w)$, but rather use plants with interval uncertainty; however, this is acceptable as the plants still satisfy a norm-bound of the form $\|\Delta H\|_{\infty^w} < \epsilon_H$). The external disturbances satisfy $\|\cdot\|_{l_{2^w}} < \alpha = 0.1$. In E_1, E_2, H_1 and H_2 of (7.46), (7.47) and (7.48), ϵ is fixed at 1. Simulation tests were performed for the case of external disturbances alone and for the case of both external disturbances together with model uncertainty. For the latter case we used the augmented system, which is given in (7.49), (7.50), (7.51) with Figure 7.6. The controller was designed based on Theorem 7.6. To check the performance of the suggested algebraic method, we also did simulations using ILC controllers designed using the MATLAB® *dhinf* command and using a FOILC learning gain matrix. For the FOILC case, we used both Arimoto-like learning gains (tuned manually) and simply the inverse of the nominal plant (i.e., $C(w) = H^{-1}$). Simulation tests for the case of external disturbances with no model uncertainty are shown in Figure 7.7 and Figure 7.8. These figures show the result of simulating the different ILC controllers with the nominal plant, for fixed l_{2^w} signals $d_I(w)$ and $d_o(w)$. Note, though these signals are iteration-varying, from the perspective of the space l_{2^w} they are fixed over the course of a simulation test. Figure 7.7 shows the ILC performance when $C(w)$ is designed from the

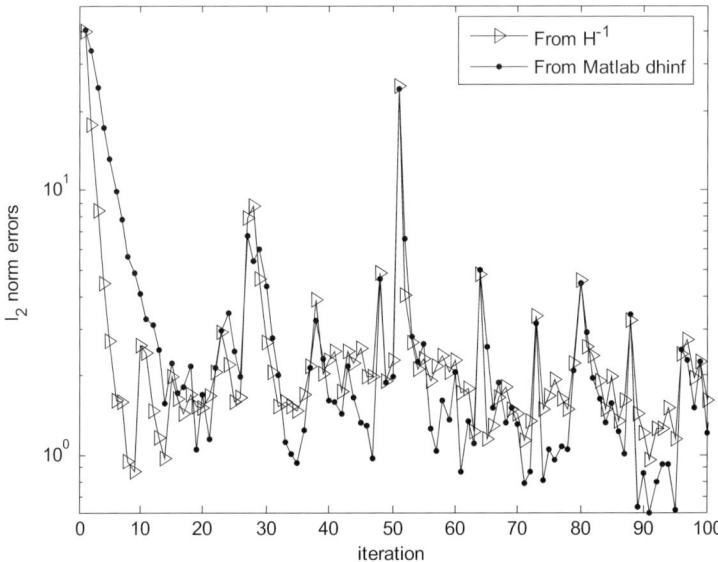

Fig. 7.7. H_∞ ILC test results with the learning gain matrices designed from the MATLAB® command *dhinf* and H^{-1}

MATLAB® *dhinf* command and from the inverse of H along the iteration axis. Figure 7.8 shows the results from the suggested algebraic method and from Arimoto-like gains. From these figures, it is not clear how to compare the performance differences between the different controller options. Note that the error does not go to zero as the number of iterations increases. That is due to the fact that the design goal of H_∞ ILC is to minimize the gain between the disturbances and the error. By assumption it cannot be driven to zero (though, the error between Y_d and the output is in fact going to zero). But, the error will depend on the signals d_I and d_o. Thus, to compare the controllers it is useful to consider the actual gain between the disturbances and the error. These are shown in Figure 7.9 (Arimoto-like gains), Figure 7.10 ($\Gamma = H^{-1}$), Figure 7.11 (from the MATLAB® *dhinf*) command, and Figure 7.12 (algebraic H_∞) for a number of different plants defined by the uncertainty model. Recall, the signal norms of interest and the gain are computed over iterations. Thus, for a given simulation (a single plant and a single set of l_{2^w} disturbance signals) we get a single number defining the norm of the error as well as the norm of the disturbance. Figures 7.9 to 7.12 plot these norms and their ratios for different plants (chosen randomly from the uncertainty model). Also recall that the algebraic H_∞ controller and the MATLAB®-based H_∞ controller were designed to achieve $\|T_{EW}\|_\infty{}^w < 1$, but the Arimoto-like gains and the inverse of H do not guarantee $\|T_{EW}\|_\infty{}^w < 1$. Thus, in these figures, we plot $\|E_k\|_{2^w}$ and $\|W_k\|_{2^w}$ as well as their ratio for 50 plants. From Figure 7.9 and Figure 7.10 (Arimoto and H^{-1}), we observe that the robust performance

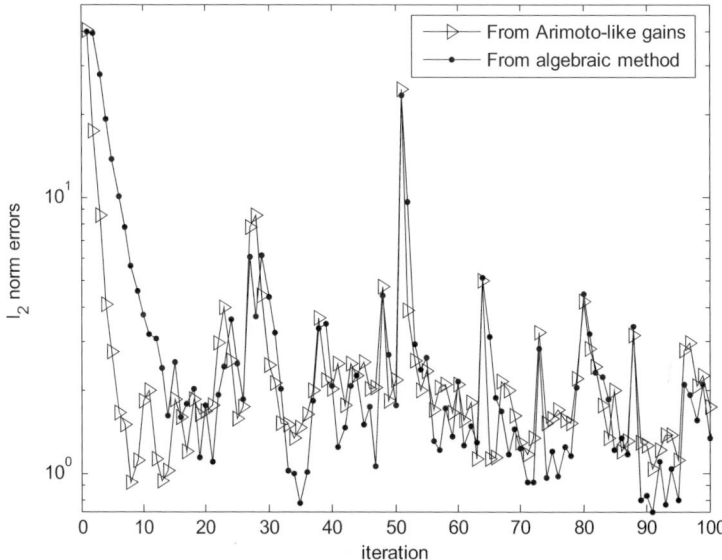

Fig. 7.8. H_∞ ILC test results with the learning gain matrices designed from the algebraic method and Arimoto-like gains

requirement $\|T_{EW}\|_{\infty^w} < 1$ is not achieved for many cases. However, from Figure 7.9 and Figure 7.10 (MATLAB® *dhinf* and algebraic method), we observe that $\|T_{EW}\|_{\infty^w} < 1$ is achieved for most plants (with exceptions due to ΔH, as noted). Clearly, from these figures, it is conclusive that the ILC system designed based on H_∞ methods are more robust than the first-order ILC systems.

7.5 Chapter Summary

In this chapter, we designed a higher-order ILC algorithm using an H_∞ framework in the iteration domain. Chapter 4, Chapter 5, and Chapter 6 considered design to guarantee monotonic convergence (MC) in the iteration axis for interval uncertainty in the time domain, while in this chapter we considered the uncertainty under the H_∞-norm topology in the iteration domain. The advantage of using the methods proposed in Chapter 4, Chapter 5, and Chapter 6 is that the MC of uncertain ILC systems can be achieved, whereas the advantage of using the method proposed in this chapter is that it can consider iteration-varying model uncertainty and external disturbances at the same time. However, there is no way to guarantee MC in the proposed H_∞ ILC framework. Hence, each method has own advantages and disadvantages, which shows that the methods proposed so far in this monograph comple-

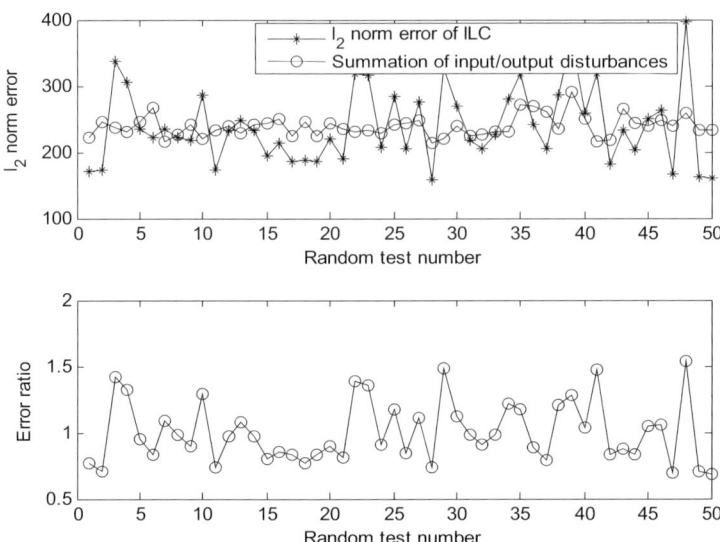

Fig. 7.9. The summation of the plant input and output disturbances $((\alpha I)d_I$ and $(\alpha I)d_o)$ and $\|(T_{EW})W_k\|_{2^w}$ with the learning gain matrix designed from Arimoto-like gains

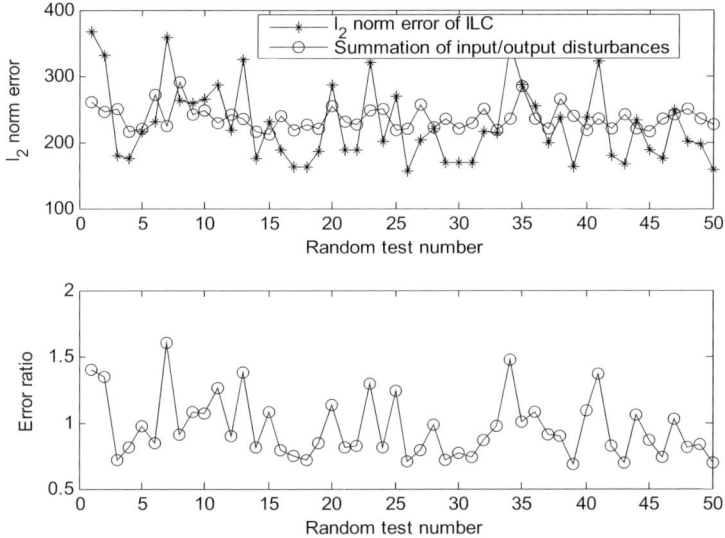

Fig. 7.10. The summation of the plant input and output disturbances $((\alpha I)d_I$ and $(\alpha I)d_o)$ and $\|(T_{EW})W_k\|_{2^w}$ with the learning gain matrix $\Gamma = H^{-1}$

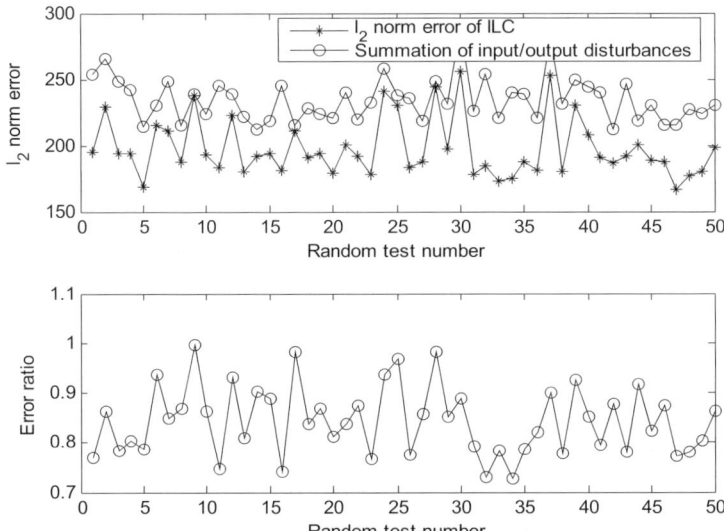

Fig. 7.11. The summation of the plant input and output disturbances $((\alpha I)d_I$ and $(\alpha I)d_o)$ and $\|(T_{EW})W_k\|_{2^w}$ with the learning gain matrix designed from the MATLAB$^{\circledR}$ *dhinf* command

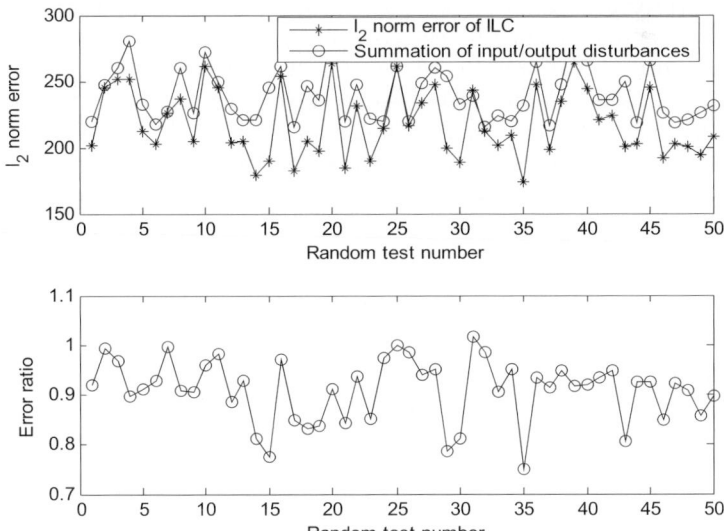

Fig. 7.12. The summation of the plant input and output disturbances $((\alpha I)d_I$ and $(\alpha I)d_o)$ and $\|(T_{EW})W_k\|_{2^w}$ with the learning gain matrix designed from the algebraic method

ment each other. The main theoretical developments of this chapter can be summarized as follows:

- First, we developed a new ILC framework using H_∞ theory for handling iteration-varying model uncertainty, external disturbances, and noises. Even though H_∞ ILC was already designed in [102, 103, 321] in the time-domain, from the fact that the time domain in ILC is finite-horizon, the results presented in these publications are restrictive from a theoretical perspective. In other words, the frequency-domain-based analysis (transformed from the finite time horizon) of [102, 103, 321] cannot consider the whole frequency range. Thus, the main research goal of this chapter is to overcome the drawback of traditional H_∞ ILC theories. For this purpose, in this chapter, the H_∞ design was performed using the discrete frequency domain associated with the iteration axis (not the time axis), under the super-vector framework, which is theoretically more sound than existing results.

- Second, for the development of a unified robust ILC framework for handling all possible robustness issues, the super-vector-based input/output relationship is cast into the traditional H_∞ framework. Then, by comparing the block diagram of the super-vector ILC with a block diagram of traditional H_∞ theory, we develop H_∞ ILC in the iteration domain.

The main contribution of this chapter is to provide a novel and unique iteration-domain, frequency-based robust ILC synthesis method that is mathematically correct. We would note that the suggested H_∞ ILC approach can be effectively used to reduce the baseline error, but to do so the filters W_i and W_o in Figure 7.5 and Figure 7.6 should be designed properly. This is a topic for further research. Similarly, we noted that the resulting controller, as in any H_∞-based controller design, has a very high order. Another future research topic is the question of how to achieve controller order reduction for this design.

8

Robust Iterative Learning Control: Stochastic Approaches

Although the algebraic H_∞ ILC approach proposed in Chapter 7 provided a unified framework for studying robust super-vector ILC (SVILC) when the plant is subject to iteration-domain uncertainty, monotonic convergence (MC) could not be guaranteed and the baseline error was not analytically established. This chapter focuses on the baseline error of robust ILC processes. Further, in our study so far in this monograph, stochastic noise has not been considered; hence, this chapter also studies the robustness problem of SVILC when the system is subject to stochastic noise.

First, a Kalman filter is utilized for finding the baseline error analytically, which enables us to design the learning gain matrix a priori. Assuming knowledge of the measurement noise and process noise statistics, the suggested ILC scheme uses a Kalman filter to estimate the error of the output measurement and a fixed learning gain controller is used to ensure that the actual error is less than a specified upper bound. An algebraic Riccati equation is solved analytically to find the steady-state covariance matrix and to prove that the system eventually converges to the baseline error.

As the second development in the chapter, iteration-varying model uncertainty and stochastic measurement noise on the iteration axis are considered in the SVILC framework. In both cases, the possibility of MC is discussed. However, it should be emphasized that in Section 8.1, an iteration-invariant learning gain matrix is used, while in Section 8.2, an iteration-variant learning gain matrix is used. Due to the fact that the baseline error is analytically found using the SVILC framework, the suggested methods in this chapter are improvements over existing stochastic ILC results [396, 395, 313, 55, 259, 257, 323], which focus on time-domain analysis.

Finally, we consider the problem of intermittent ILC, where we assume that on each trial some measurement data is lost (stochastically). We show how a Kalman filter-based approach in the iteration domain can compensate for the loss of data. This is a useful result for the problem of ILC in a networked control system environment.

8.1 Baseline Error Estimation in the Iteration Domain

Recently, in ILC research, the idea of a fundamental baseline error has been pointed out [64] and an effort has been made to resolve this problem. In detail, in [64] a model-based approach was used to reduce the baseline error of the deterministic ILC process. However, applications of the results of [64] are limited because the results are based on the internal model principle. In this section, an alternate approach to reduce the baseline error of noise-driven ILC systems is proposed using a Kalman filter. Although Kalman filter-based ILC design methods have been suggested in the existing literature [396, 395, 313, 55, 259, 257, 323], their main stability analysis was limited to the time domain and all result in an iteration-dependent learning gain design. For example, in [396], stochastic ILC was proposed based on a full state model and iteration-varying learning gains, even though the assumption of knowledge of the A matrix was removed in [395]. Several works have proposed iteration-domain filtering. In [259, 257] convergence was analyzed using quadratic programming and in [323] an iteration-varying learning gain matrix was used to estimate the output Y_k. Both of these works assumed that the input–output relation matrix H (i.e., $Y_k = HU_k$ where U_k is the input and Y_k is the output) is available and they also used iteration-varying learning gain matrices. Recently, in [55], nonlinear stochastic ILC was proposed, where the nonlinear system model was considered to be unknown. However, they also used iteration-varying gains and the slow convergence speed was observed as a disadvantage.

In this section, we take a different approach by focusing on estimating the error E_k on the iteration axis with an iteration-domain Kalman filter and using this estimate in the ILC-type update algorithm. Estimating E_k enables us to design the fixed learning gain matrix, which need not be updated every iteration. Due to the fact that the fixed learning gain matrix Γ is chosen to guarantee the convergence of stochastic ILC system, the approach presented in this section is clearly different from existing works and could be considered more practical. However, in this monograph we still assume that the H matrix is available and the statistics of the noise are known, as done in [396, 313, 259, 257, 323]. This kind of assumption seems to be a betrayal of the main reason for using ILC, because ILC was initially developed to achieve the perfect tracking for unknown model control systems [28]. However, note that the advantage of using ILC is not only in the case of model uncertainty. For example, ILC has been proven to be very effective in controlling non-minimum phase systems; ILC can improve convergence speed and give perfect tracking when a standard controller cannot, even when the model is known; and the monotone (without overshoot) convergence can be guaranteed by ILC, which is a practically important issue as commented in Chapter 1.

The main focus of this section is to find the baseline error analytically, which is dependent on the designed learning gain matrix, in addition to, under some conditions, guaranteeing the monotonic convergence of the stochastic ILC system along the iteration domain. Hence, even though this section assumes the H matrix is known, as done in other ILC results, the contributions of this section over existing works can be summarized as: (i) the steady-state estimate of the error on the iteration axis can be computed analytically, (ii) a fixed learning gain matrix is used and with this fixed gain matrix, the baseline error can be significantly reduced (in a stochastic sense), and (iii) under some conditions, the monotonic convergence of ILC system (in a stochastic sense) can be guaranteed until it reaches a baseline error that can be calculated in advance.

8.1.1 Modeling

In this section, it will be shown that the SVILC scheme can be well-cast into the discrete linear Kalman filter framework.

Standard Kalman Filtering

Let us start by introducing the standard Kalman filter algorithm. The following discrete-time stochastic system is considered [153]:

$$x_{k+1} = Ax_k + Gv_k \tag{8.1}$$
$$\tilde{z}_k = Cx_k + w_k \tag{8.2}$$

where $v_k \sim N(0, Q)$ is zero-mean Gaussian process noise with covariance Q; $w_k \sim N(0, R)$ is zero-mean Gaussian measurement noise with covariance R; $x_k \in \mathcal{R}^n$ is the state; $A \in \mathcal{R}^{n \times n}$; $C \in \mathcal{R}^{n \times n}$; $\tilde{z}_k \in \mathcal{R}^n$ is the actual measurement; $G \in \mathcal{R}^{n \times n}$, and k is the discrete-time index (not, in this case, the trial or iteration index). Then, the discrete linear Kalman filter is designed as

- Propagation:

$$\bar{x}_k = A\hat{x}_{k-1} \tag{8.3}$$
$$\bar{P}_k = A\hat{P}_{k-1}A^T + GQG^T \tag{8.4}$$

- Correction:

$$K_k = \bar{P}_k C^T (C\bar{P}_k C^T + R)^{-1} \tag{8.5}$$
$$\hat{x}_k = \bar{x}_k + K_k(\tilde{z}_k - C\bar{x}_k) \tag{8.6}$$
$$\hat{P}_k = (I - K_k C)\bar{P}_k, \tag{8.7}$$

where P_k is the covariance of the estimated state error, K_k is the Kalman gain matrix, ¯ stands for the propagated state, and ˆ stands for the corrected state using the Kalman gain matrix.

Stochastic SVILC Frameworks

Now it will be shown that SVILC can be designed based on the above discrete Kalman filter structure along the iteration axis. As explained in Section 4.1, SVILC is updated by the following mechanism:

$$U_{k+1} = U_k + \Gamma E_k \tag{8.8}$$

$$Y_k = HU_k, \tag{8.9}$$

where $U_k \in \mathcal{R}^n$ is the control input at the kth iteration; $\Gamma \in \mathcal{R}^{n \times n}$ is the fixed learning gain matrix; $E_k \in \mathcal{R}^n$ is the error vector at the kth iteration, which is calculated by $E_k = Y_d - Y_k$; $H \in \mathcal{R}^{n \times n}$ is the system Markov matrix, and $Y_k \in \mathcal{R}^n$ is the measured output. Notice that in (8.1)–(8.7), k represents the time point, but in (8.8)–(8.9), k represents the iteration axis. As we have described, it is a standard lifting technique in ILC to transform a system operated repetitively as per the ILC paradigm into a multivariable system of the form (8.1)–(8.7), where the iteration index can be viewed as a discrete-time index.

To continue, noting that $Y_{k+1} = HU_{k+1}$ and defining $\Delta_k^u \equiv U_{k+1} - U_k$, we obtain the following relationship:

$$E_{k+1} = E_k - H\Delta_k^u. \tag{8.10}$$

Let us suppose that the measured state at the kth iteration is Y_k and during this measurement there exists measurement noise w_k with $w_k \sim N(0, R)$. Now introducing the measurement noise w_k, we have

$$\tilde{E}_k = Y_d - Y_k - w_k = E_k - w_k. \tag{8.11}$$

In (8.10), $\Delta_k^u = U_{k+1} - U_k$ and from (8.8), since the measured error is used for control update, we obtain $U_{k+1} - U_k = \Gamma \tilde{E}_k$. Let us suppose that during the control law update, there is process noise v_k' that is zero-mean Gaussian process noise with covariance Q', i.e., $v_k' \sim N(0, Q')$. Then, by writing $U_{k+1} = U_k + \Gamma \tilde{E}_k + v_k'$, the control update law is found as

$$U_{k+1} = U_k + \Gamma(E_k - w_k) + v_k'$$
$$= U_k + \Gamma E_k - \Gamma w_k + v_k'. \tag{8.12}$$

Now, writing $-\Gamma w_k + v_k' \equiv v_k$, we have $U_{k+1} = U_k + \Gamma E_k + v_k$, which yields

$$\Delta_k^u = \Gamma E_k + v_k. \tag{8.13}$$

Now, from $\mathbf{E}\left[(-\Gamma w_k + v_k')(-\Gamma w_k + v_k')^T\right] = \Gamma \mathbf{E}\left[w_k w_k^T\right]\Gamma^T + \mathbf{E}\left[v_k'(v_k')^T\right] = \Gamma R \Gamma^T + Q' = Q$, we simply consider v_k as zero-mean Gaussian process noise with covariance Q, i.e., $v_k \sim N(0, Q)$. Then, inserting (8.13) into (8.10) yields

$$E_{k+1} = E_k - H(\Gamma E_k + v_k)$$
$$= (I - H\Gamma)E_k - Hv_k, \tag{8.14}$$

which is the same form as the typical FOILC error update rule (3.9) except for the process and measurement noises. Now, from (8.11) and (8.14), the noise-driven ILC system can be formulated in a state-space form along the iteration axis as

$$E_{k+1} = AE_k + Gv_k \tag{8.15}$$
$$\tilde{E}_k = CE_k + w_k, \tag{8.16}$$

where $A = I - H\Gamma$, $G = -H$, and $C = I$. The following technical assumption is needed.

Assumption 8.1. *In (8.11)–(8.16), it is assumed that the process and measurement noises v'_k and w_k are zero-mean white Gaussian noises on the iteration axis. Furthermore, the noise characteristics R and Q' are assumed to be known (so, Q is also known from $Q = \Gamma R \Gamma^T + Q'$).*

Remark 8.2. It is worthwhile to note that the measurement noise w_k may come from the initial state reset error. When $x_k(0)$ is not fixed,[1] (8.9) should be changed to

$$Y_k = HU_k + Wx_k(0), \tag{8.17}$$

where $x_k(0)$ is the initial state and $W = [v_1^T, v_2^T, \ldots, v_{n+1}^T]^T$ with $v_i = CA^i$. Then, we can write $x_k(0) = \bar{x}_k(0) + \Delta x_k(0)$ where $\bar{x}_k(0)$ is the nominal initial state and $\Delta x_k(0)$ is the initial state error. Then, as done in [397], assuming zero-mean Gaussian $\Delta x_k(0)$, simply we can equalize $Wx_k(0) = w_k$. Thus, the stochastic SVILC frame proposed in this monograph can effectively handle the initial state error problem.

Next, by comparing (8.15)–(8.16) to (8.1)–(8.2), and matching $A = I - H\Gamma$, $x_k = E_k$, $G = -H$, $\tilde{z}_k = \tilde{E}_k$ and $C = I$, we can derive the following Kalman-filter-based recursive update formula for stochastic SVILC, which can be used for the baseline error estimation:

$$\bar{E}_k = (I - H\Gamma)\hat{E}_{k-1} \tag{8.18}$$
$$\bar{P}_k = \hat{P}_{k-1} + HQH^T \tag{8.19}$$
$$K_k = \bar{P}_k(\bar{P}_k + R)^{-1} \tag{8.20}$$
$$\hat{E}_k = \bar{E}_k + K_k(\tilde{z}_k - \bar{E}_k) \tag{8.21}$$
$$\hat{P}_k = (I - K_k)\bar{P}_k, \tag{8.22}$$

where $\tilde{z}_k = \tilde{E}_k$. Observe that in the algorithm (8.18)–(8.22), the inputs are $\hat{E}_{k-1}$ and $\tilde{z}_k$, and the output is $\hat{E}_k$.

---

[1] In ILC, as written in (1.20), generally it is required that the initial state is fixed at the same place. However, in practice, it could be different each iteration.

8.1.2 Analytical Solutions

In this section, it will be shown that from the developed algorithm the baseline error of the SVILC system along the iteration axis can be estimated. To obtain this result we need to show that $\hat{P}_k$ converges as $k \to \infty$. For this purpose, the following lemmas are developed first.

Lemma 8.3. *If the covariance matrix of the process noise vector is given as $Q_{ii} \neq 0$ if $i = j$, $Q_{ij} = 0$, if $i \neq j$ and CB is full rank, then HQH^T is nonsingular.* ∎

Proof. If CB is full rank, H is nonsingular and Q is positive definite. Thus, from $\det(HQH^T) = \det H \cdot \det Q \cdot \det H^T$, since $\det H \neq 0$ and $\det Q \neq 0$, we have $\det(HQH^T) \neq 0$. Therefore, HQH^T is nonsingular. ∎

Lemma 8.4. *Under the same conditions as Lemma 8.3, if A is any covariance matrix, then $A + HQH^T$ is positive definite (p.d.), i.e., $A + HQH^T > 0$.* ∎

Proof. Since $A + HQH^T$ is symmetric, if $x^T(A + HQH^T)x > 0$ for all nonzero vectors x, then $A + HQH^T$ is positive. Since any covariance matrix is positive or at least semi-positive definite, $x^T A x \geq 0$. Therefore, if $x^T(A + HQH^T)x > 0$, then $A + HQH^T$ is positive definite. Let us make the following relationship:

$$x^T(A + HQH^T)x = x^T A x + x^T HQH^T x = x^T A x + y^T Q y,$$

where we used $y = H^T x$. Here noticing $Q > 0$, because it is diagonal matrix with zero off-diagonal terms, $y^T Q y > 0$ for all nonzero y. However, the nonzero condition $y \neq 0$ is enforced. Now, we have to prove that $y \neq 0$ if and only if $x \neq 0$. It is easy to see that the column vectors and row vectors are linearly independent because the H matrix is nonsingular. Hence, from

$$H^T x = p_1 x_1 + p_2 x_2 + \cdots + p_n x_n,$$

where $[p_1, p_2, \ldots, p_n] = H^T$ and $[x_1, x_2, \ldots, x_n]^T = x$, we know that only the trivial solution $x_1 = x_2 = \cdots = x_n = 0$ makes $H^T x = 0$. Therefore, we know that for all nonzero y, $y^T Q y > 0$. This completes the proof. ∎

Now, using the lemmas given above, let us develop the following theorem.

Theorem 8.5. *If there exists a solution X for the following algebraic Riccati equation (ARE):*

$$AX + XA - XBX + C = 0 \tag{8.23}$$

with $A = \frac{1}{2}I$, $B = (HQH^T)^{-1}$ and $C = R$, then $\lim_{k \to \infty} \hat{P}_k = \hat{P}^* \equiv X - HQH^T$. ∎

Proof. Inserting (8.19) into (8.20) yields

$$K_k = (\hat{P}_{k-1} + HQH^T)(\hat{P}_{k-1} + HQH^T + R)^{-1}. \tag{8.24}$$

Also, inserting (8.19) and (8.24) into (8.22), we obtain

$$
\begin{aligned}
\hat{P}_k &= \left(I - (\hat{P}_{k-1} + HQH^T)(\hat{P}_{k-1} + HQH^T + R)^{-1} \right) (\hat{P}_{k-1} + HQH^T) \\
&= \hat{P}_{k-1} + HQH^T - (\hat{P}_{k-1} + HQH^T)(\hat{P}_{k-1} + HQH^T + R)^{-1} \\
&\quad \times (\hat{P}_{k-1} + HQH^T).
\end{aligned}
$$

Now, since $\hat{P}_k = \hat{P}_{k-1}$ is equivalent to the convergence of $\hat{P}_k$, if and only if

$$
HQH^T - (\hat{P}_{k-1} + HQH^T)(\hat{P}_{k-1} + HQH^T + R)^{-1}(\hat{P}_{k-1} + HQH^T) = 0,
\tag{8.25}
$$

then $\hat{P}_k$ converges to $\lim_{k \to \infty} \hat{P}_k = \hat{P}^*$. For convenience, by writing $X \equiv \hat{P}_{k-1} + HQH^T$, we have

$$
HQH^T - X(X + R)^{-1}X = 0.
\tag{8.26}
$$

From Lemma 8.4, since X is nonsingular, we can change (8.26) to:

$$
HQH^T X^{-1} = X(X + R)^{-1}.
\tag{8.27}
$$

Next, using Lemma 8.3 and taking the inverse of both sides of (8.27), we obtain:

$$
X(HQH^T)^{-1} = (X + R)X^{-1}.
\tag{8.28}
$$

Thus, we have

$$
\begin{aligned}
X(HQH^T)^{-1}X = (X + R) &\Leftrightarrow X - X(HQH^T)^{-1}X + R = 0 \\
&\Leftrightarrow \frac{I}{2}X + X\frac{I}{2} - X(HQH^T)^{-1}X + R = 0.
\end{aligned}
\tag{8.29}
$$

Now, (8.29) is a typical algebraic Riccati equation, and $A = \frac{1}{2}I$, $B = (HQH^T)^{-1}$ and $C = R$. Thus, if there exists X such that (8.29) is satisfied, then $P_k = P_{k-1}$. This completes the proof. ■

In the next theorem, we show that there always exists a unique solution of the ARE of Theorem 8.5. For this proof, we use a well-known existence condition for continuous AREs, which is summarized in the following lemma.

Lemma 8.6. *[107] For the following ARE:*

$$
A^T K + KA - KBB^T K + Q = 0
\tag{8.30}
$$

with $Q = C^T C$, if (A,B) is stabilizable and (A,C) is observable, then there exists a unique $K = K^T > 0$ satisfying (8.30). ■

Theorem 8.7. *In stochastic SVILC, if CB is full rank, then there always exists a unique p.d solution X of (8.23).*   ∎

Proof. Let us define matrices

$$V_{ii} = \frac{1}{\sqrt{Q_{ii}}}, \ V_{ij} = 0, \text{when } i \neq j, \ U_{ii} = \sqrt{Q_{ii}}, \ U_{ij} = 0, \text{when } i \neq j$$

$$\overline{A} := \frac{1}{2}I; \ \overline{B} := (H^{-1})^T V; \ \overline{C} := U.$$

Then, we know that there always exists a matrix $\overline{K}$ such that $\lambda(\overline{A} + \overline{B}\overline{K}) < 0$, because $\overline{B}$ is invertible. Also since $\overline{A}$ and $\overline{C}$ are nonzero diagonal matrices, the pair $(\overline{A}, \overline{C})$ is observable. Therefore, by Lemma 8.6, there exists a positive definite solution X.   ∎

Remark 8.8. From [486], in addition to the conditions of Lemma 8.6, if (A,B) is controllable, then we always have $\lambda(A - BB^T K) < 0$. Since $(\overline{A}, \overline{B})$ is also controllable, we have

$$\lambda\left(1/2I - (HQH^T)^{-1}X\right) < 0. \tag{8.31}$$

Then, using (8.31), we can find an analytical condition for $X - HQH^T > 0$. This work is a topic for future work.

The condition for convergence is always an important consideration in ILC. The convergence condition of the stochastic SVILC system can be established based on this convergence condition, the baseline error of the stochastic SVILC can be subsequently estimated. Furthermore, under more restrictive conditions it can be shown that as a special case the convergence could be monotonic. In the remainder of this section we first give a theorem that provides an upper bound for the baseline error and we then give a result that establishes conditions for MC. Before further proceeding, for the analytical solution, we need a definition with regard to the norm of a noise vector.

Definition 8.9. *The stochastic norm of the measurement noise w_k, ($w_k \sim N(0, R)$), a zero-mean white Gaussian process, is defined as*

$$\|w_k\| = \sqrt{\sum_{i=1}^{n} R_{ii}},$$

where R_{ii} represents the diagonal term of the noise covariance matrix.[2]   ∎

---

[2] In fact, it is not possible to define the norm of a Gaussian noise vector (denoted by w), because the l_2-norm or the l_∞-norm could be infinity. Thus, we cannot say that a random vector is in l_p space. However, $E(\|w\|) < \infty$. Thus, using the expected values of the measurement noise vector enables us to bound the norm of a noise vector stochastically. This assumption is practically acceptable due to Remark 8.2.

In the following theorems, when we use the terminology *in a stochastic sense*, it implies that stochastic norms are used as defined above. Also, note that the symbol $\| \cdot \|$ is used to denote the l_2-norm.

Theorem 8.10. *Defining* $S^{P^*} \equiv \sqrt{\sum_{i=1}^{n}(\hat{P}^*)_{ii}}$, *if* $\|I - H\Gamma\| < 1$, *the estimated baseline error converges, in a stochastic sense, within an upper bound given by:*

$$\|\hat{E}^*\| \leq \frac{1}{1 - \|I - H\Gamma\|} \left\| (I - K^*)^{-1} K^* \right\| \left(S^{P^*} + \|w_k\| \right), \qquad (8.32)$$

where $K^* \equiv \bar{P}^*(\bar{P}^* + R)^{-1}$ *and* $\bar{P}^* = \hat{P}^* + HQH^T$. ∎

Proof. From Theorem 8.5, we know there exists $\hat{P}^*$ as $k \to \infty$. Thus, from (8.19)–(8.20), we obtain the steady-state values of $\bar{P}_k$ and K_k to be $\bar{P}^* = \hat{P}^* + HQH^T$ and $K^* = \bar{P}^*(\bar{P}^* + R)^{-1}$. Now, substituting K^* and (8.18) into (8.21), we have:

$$\begin{aligned} \hat{E}_k &= (I - H\Gamma)\hat{E}_{k-1} + K^*(\tilde{z}_k - \hat{E}_{k-1} + H\Gamma\hat{E}_{k-1}) \\ &= (I - K^*)(I - H\Gamma)\hat{E}_{k-1} + K^*\tilde{z}_k. \end{aligned} \qquad (8.33)$$

From the relationships $\tilde{z}_k = Y_d - \tilde{Y}_k = Y_d - Y_k + w_k$ and $Y_d - Y_k = E_k$, and from the definition $\Delta E_k = E_k - \hat{E}_k$, we also obtain $\tilde{z}_k = \Delta E_k + \hat{E}_k + w_k$. Therefore, (8.33) becomes:

$$\hat{E}_k = (I - K^*)(I - H\Gamma)\hat{E}_{k-1} + K^*(\Delta E_k + \hat{E}_k + w_k). \qquad (8.34)$$

Thus, we have

$$\hat{E}_k = (I - H\Gamma)\hat{E}_{k-1} + (I - K^*)^{-1}K^*(\Delta E_k + w_k) \qquad (8.35)$$

and hence, from the general solution of the first-order discrete-time system, as $k \to \infty$, if $\|I - H\Gamma\| < 1$, we then obtain the inequalities:

$$\begin{aligned} \|\hat{E}^*\| &\leq \left\| \frac{1}{1 - \|I - H\Gamma\|}(I - K^*)^{-1}K^* \right\| \|(\Delta E_k + w_k)\| \\ &\leq \frac{1}{1 - \|I - H\Gamma\|} \left\| (I - K^*)^{-1}K^* \right\| (\|\Delta E_k\| + \|w_k\|), \end{aligned}$$

where we used $\left\| \frac{1}{1-\|I-H\Gamma\|} \right\| = \frac{1}{1-\|I-H\Gamma\|}$, because $\|I - H\Gamma\| < 1$. Now, since $\|\Delta E_k\| = \sqrt{\sum_{i=1}^{n} \Delta E_k(i)^2}$ and $\hat{P}^* = E\{\Delta E_k^*(\Delta E_k^*)^T\}$ where ΔE_k^* is an estimated steady-state error boundary, we obtain $\sqrt{\sum_{i=1}^{n}(\hat{P}^*)_{ii}} = \sqrt{\sum_{i=1}^{n} \Delta E_k^*(i)^2}$, because ΔE_k^* is in the steady state. Therefore,

$$\|\hat{E}^*\| \leq \frac{1}{1 - \|I - H\Gamma\|} \left\| (I - K^*)^{-1}K^* \right\| \left(S^{P^*} + \|w_k\| \right). \qquad (8.36)$$

This completes the proof. ∎

Theorem 8.10 shows that as $K^* \to 0$, $\|\hat{E}^*\| \to 0$. This means that from $\bar{P}^* = \hat{P}^* + HQH^T$ and $K^* = \bar{P}^*(\bar{P}^* + R)^{-1}$, as $\hat{P}^* \to 0$ and $Q \to 0$, the baseline error goes to zero. That is, in the noise-free measurement case it is always possible to have ILC convergence with zero error, but in the more practical noisy-measurement setting, ILC must always live with a nonzero baseline error, as pointed out in [64].

Remark 8.11. Theorem 8.10 provides an upper bound for the estimated error $\hat{E}_k$. But, although the upper bound of the actual error E_k could be different from the upper bound of $\hat{E}_k$ given in Theorem 8.10, since $\|\hat{P}^*\|$ is known and $\hat{P}_k$ bounds $E_k - \hat{E}_k$, the upper bound of the actual error can be estimated by $\|E_k\| \le \|\hat{E}_k\| + S^{P^*}$.

Next we consider conditions for monotonic convergence.

Theorem 8.12. *Let us suppose that the error covariance matrix is in the steady state, and the iteration count is at the $(k-1)^{\text{th}}$ trial. If $S^{P^*} + \|w_k\| < \|\hat{E}_{k-1}\|$ and*

$$\|I - H\Gamma\| + \|(I - K^*)^{-1}K^*\| < 1, \qquad (8.37)$$

then the estimated error vector can, in a stochastic sense, be MC, i.e., $\|\hat{E}_k\| < \|\hat{E}_{k-1}\|$. ∎

Proof. From (8.35), the following inequalities can easily be derived:

$$
\begin{aligned}
\|\hat{E}_k\| &= \|(I - H\Gamma)\hat{E}_{k-1} + (I - K^*)^{-1}K^*(\Delta E_k + w_k)\| \\
&\le \|(I - H\Gamma)\|\|\hat{E}_{k-1}\| + \|(I - K^*)^{-1}K^*\|(\|\Delta E_k\| + \|w_k\|) \\
&= \|(I - H\Gamma)\|\|\hat{E}_{k-1}\| + \|(I - K^*)^{-1}K^*\|(S^{P^*} + \|w_k\|). \quad (8.38)
\end{aligned}
$$

Therefore, if $S^{P^*} + \|w_k\| < \|\hat{E}_{k-1}\|$, we then have the inequality $\|(I - H\Gamma)\|\|\hat{E}_{k-1}\| + \|(I - K^*)^{-1}K^*\|(S^{P^*} + \|w_k\|) < (\|(I - H\Gamma)\| + \|(I - K^*)^{-1}K^*\|)\|\hat{E}_{k-1}\|$. Hence, if $\|I - H\Gamma\| + \|(I - K^*)^{-1}K^*\| < 1$, the MC condition $\|\hat{E}_k\| < \|\hat{E}_{k-1}\|$ is guaranteed. ∎

Remark 8.13. Theorem 8.12 also shows that as $K^* \to 0$, the MC condition can be easily satisfied. Further, we see that the learning gain matrix $\Gamma = H^{-1}$ is the best possible gain for MC of the stochastic SVILC system, as is the case in all first-order ILC algorithms.

Theorem 8.12 also provides an important design strategy. Let us suppose that we have designed an ILC learning gain Γ such that

$$\|I - H\Gamma\| + \|(I - K^*)^{-1}K^*\| < 1 \qquad (8.39)$$

and further suppose that the system is in the steady state. Then, at the $(k-1)^{\text{th}}$ trial, although the estimation error $\|\hat{E}_{k-1}\|$ might be very big due to unexpected noises (more than $S^{P^*} + \|w_k\| = S^{P^*} + \sqrt{\sum_{i=1}^{n} R_{ii}}$), the estimated error is forced to decrease at the kth iteration trial in a stochastic sense, which is also an expected result from Theorem 8.10. Thus, we have the following remark:

Remark 8.14. From Theorem 8.10, it was shown that $\|\hat{E}_k\|$ and $\|E_k\|$ can be upper-bounded, and from Theorem 8.12, when $\|\hat{E}_k\|$ is bigger than a specified value (but after the first convergence to the specified value is achieved), MC is enforced. But, both theorems are associated with the learning gain matrix Γ. Thus, by properly designing Γ off-line, a specified design requirement can be achieved.

We will illustrate this result in the next subsection.

8.1.3 Simulation Illustrations

In this subsection, an example is provided to illustrate the validity of the suggested Kalman filter-augmented SVILC scheme. The following discrete-time system, which was given in Section 5.3, is used:

$$x_{t+1} = \begin{bmatrix} -0.50 & 0.00 & 0.00 \\ 1.00 & 1.24 & -0.87 \\ 0.00 & 0.87 & 0.00 \end{bmatrix} x_t + \begin{bmatrix} 1.0 \\ 0.0 \\ 0.0 \end{bmatrix} u_t \qquad (8.40)$$

$$y_t = [2.0 \quad 2.6 \quad -2.8] x_t, \qquad (8.41)$$

The system has poles at $[0.62 + j0.62, 0.62 - j0.62, -0.50]$ and zeros at $[0.65, -0.71]$. In this test, we assume a zero initial condition on every trial and 10 discrete-time points. The desired repetitive reference trajectory is $Y_d(j) = 5\sin(8.0(j-1)/10)$, $j = 1, \ldots, 10$. The same learning gain matrix Γ as used Section 5.3 is used, which was determined by an optimization method based on Lyapunov stability analysis. For random Gaussian noise, the MATLAB® command *randn* is used, with the covariance of v'_k taken as 10^{-4} and the covariance of w_k taken as 1. That is, diag$(Q') = 10^{-4}$ and diag$(R) = 1.0$, because practically it is reasonable to assume that Q' is very small. Figures 8.1 to 8.4 show the test results. Figure 8.1 shows the norms of the tracking error $E_k = Y_d - Y_k$ with respect to iteration number for cases with and without a Kalman filter along the iteration axis. The solid line is the calculated upper boundary of E_k. Figure 8.2 shows the norms of $\hat{E}_k$ and the calculated upper boundary of $\hat{E}_k$ with respect to the iteration number. As shown in Figure 8.1, with the Kalman filter, the baseline error is significantly reduced. Figure 8.2 also shows the calculated upper bound of the estimated baseline error (dash-dot line, 0.9088) from Theorem 8.10. It can be observed that the actual baseline error is well-bounded by the previously calculated upper-boundary values. Figure 8.3 shows the covariance matrix of the estimated error vector. The top left plot is the 3-D representation of the actual P_k (P_k is a 10×10 matrix). The top right plot is the 3-D representation of the estimated P_k (P_k^*). The bottom left plot is the 3-D representation of $P_k - P_k^*$. The bottom right (top) plot is the norm of the covariance of the estimated error $\|\hat{P}_k\|$ with respect to iteration number and the bottom right (bottom) plot is $\|\hat{P}_k - P_k^*\|$. As shown in these figures, $\hat{P}_k$ converges to the

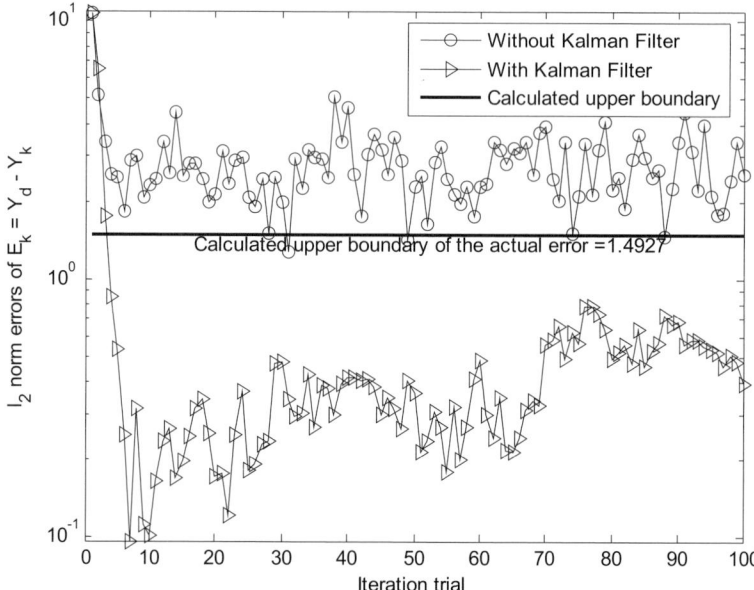

Fig. 8.1. Norms of tracking error $E_k = Y_d - Y_k$ w.r.t. iteration number for cases with and without KF in iteration axis. The solid line is the calculated upper boundary of E_k.

previously calculated P_k^* accurately. In particular, the bottom right figures show $\hat{P}_k$ quantitatively, because we used $\|\hat{P}_k\|$. As shown in these figures, as the iteration number increases, the error of the estimated state (in this monograph, $\Delta E_k = E_k - \hat{E}_k$) decreases and after the 15th iteration, there is a steady-state baseline error. As shown in the bottom figure, after the 15th iteration, $\|\hat{P}_k - P_k^*\|$ is almost zero. This means that the previously estimated P_k^* is very accurate and reliable. Figure 8.4 shows the estimated error $\hat{E}_k$ and actual error E_k with respect to time and iteration. The top left plot is the 3-D representation of actual E_k. The top right plot is the 3-D representation of the actual $\hat{E}_k$. The bottom plot is the 3-D representation of $E_k - \hat{E}_k$. As shown in these figures, the suggested algorithm estimates E_k reliably even though it is not perfect. The bottom difference corresponds to the difference between Figure 8.1 and Figure 8.2.

8.1.4 Concluding Remarks

In this section, a new Kalman filter scheme for the SVILC system was developed. The new method provides an effective ILC design scheme for systems with measurement noise with little extra implementation cost. Through a numerical example, the validity of the newly proposed method was illustrated.

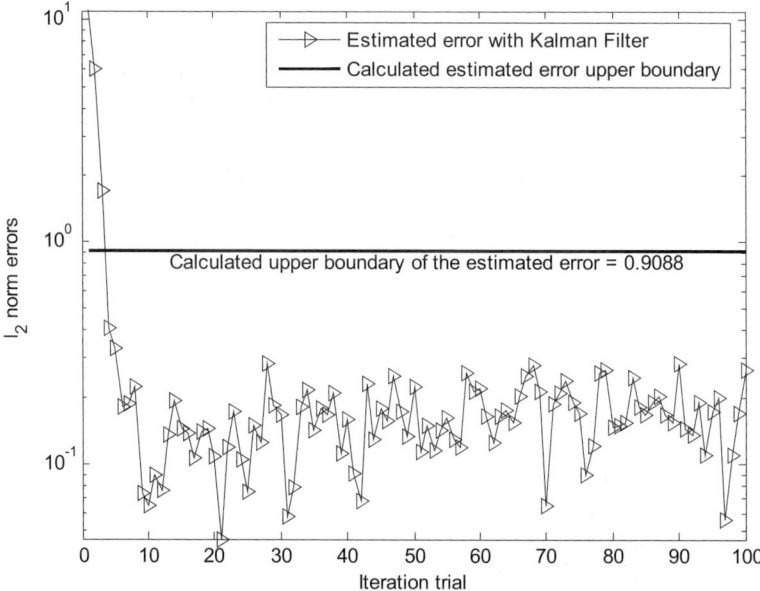

Fig. 8.2. Norms of $\hat{E}_k$ and the calculated upper boundary of $\hat{E}_k$ w.r.t. iteration number

The key idea of the new method is that the learning gain matrix can be determined a priori (not iteration-varying), but the output error is estimated on-line. As a main contribution of this section, it was shown that the baseline error of uncertain ILC process could be significantly reduced by the suggested stochastic ILC scheme and the upper bound of the estimated error (and actual error) can be calculated a priori, given the input and measurement noise covariances. From the fact that the new algorithm computes the learning gain matrix off-line, the work proposed in this section is significantly different from the existing stochastic ILC algorithms. Furthermore, in Remark 8.2, we discussed the possibility of using the suggested scheme to handle the initial reset problem.

Finally, we make the following remark about how to relax the requirement of assuming knowledge of the system Markov matrix H:

Remark 8.15. Although in Theorem 8.10 the condition $\|I - H\Gamma\| < 1$ was required, this condition can be relaxed to $\rho(I - H\Gamma) < 1$, where $\rho(\cdot)$ is the spectral radius. In this case, the system will be bounded-input, bounded-output stable (BIBO stable). Note again, if we only require $\rho(I - H\Gamma) < 1$, then we do not need to know A matrix. That is, as proven in Theorem 8.7, if CB is full rank, there exists a steady-state $\hat{P}_k$ as $k \to \infty$, and also from (8.35), if CB is full rank, we can then satisfy the condition $\rho(I - H\Gamma) < 1$. Therefore, without knowing A, the suggested stochastic ILC scheme guarantees the steady-state error and the BIBO stability. However, without A, Theorem 8.12

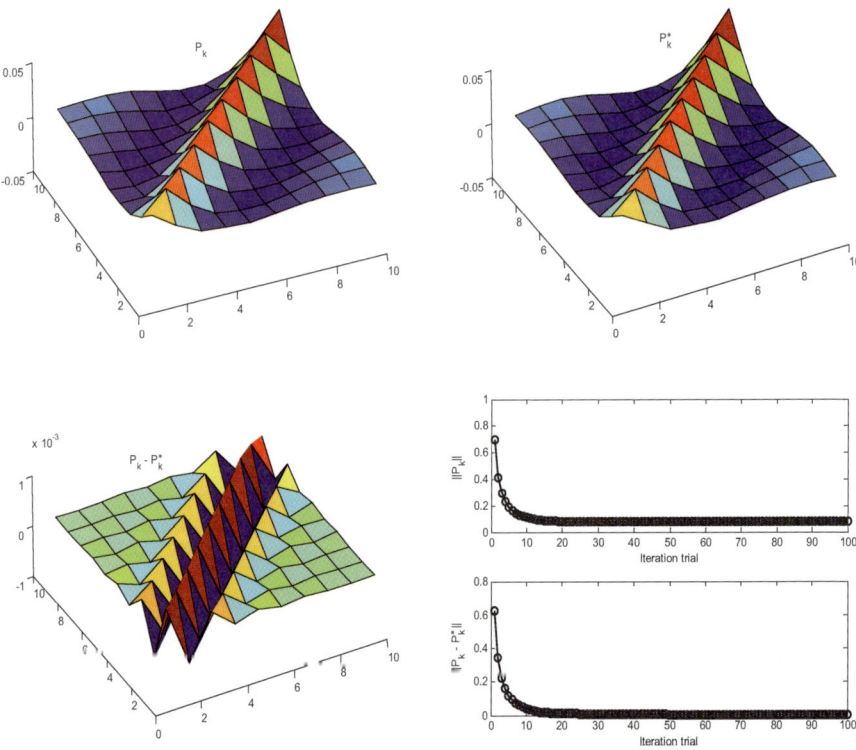

Fig. 8.3. Top left: 3-D representation of actual P_k (P_k is 10×10 matrix). Top right: 3-D representation of estimated P_k (P_k^*). Bottom left: 3-D representation of $P_k - P_k^*$. Bottom right: (Top) Norm of covariance of the estimated error $\|\hat{P}_k\|$ w.r.t. iteration number; (Bottom) $\|\hat{P}_k - P_k^*\|$.

is no longer valid and the baseline error can not be estimated analytically. Hence, the suggested algorithm can be used for two purposes: (i) with knowledge of A, the baseline error can be estimated and MC can be guaranteed, but (ii) without knowledge of A, BIBO stability and steady-state error are still achievable.

In this monograph, we assume that the Markov matrix H is known; based on this assumption, we have developed an algorithm for estimating and reducing the baseline error. Although this approach has been widely used in ILC [396, 313, 259, 257, 323] with an exception [55], in order to conform the main purpose of the ILC algorithm, we may have to consider uncertain (or unknown) H matrix. In this case, as commented in [396], we need to identify the model. For this purpose, various methods could be considered as done in [396, 323]. In this monograph, however, we propose using the Wiener filter or the least squares method for its simplicity. This is described in detail in the following section.

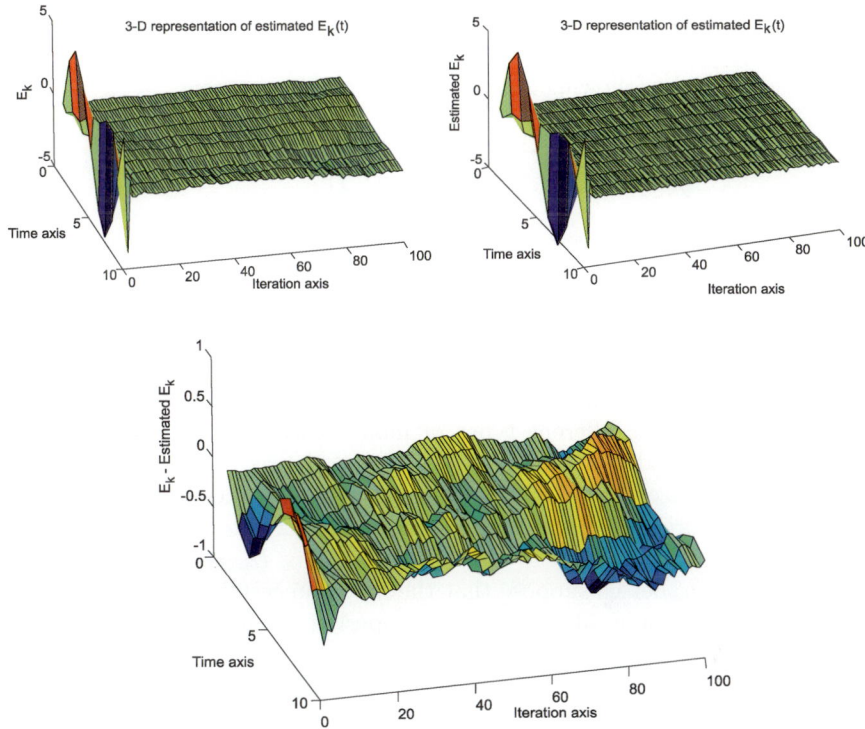

Fig. 8.4. Top left: 3-D representation of actual E_k. Top right: 3-D representation of actual $\hat{E}_k$. Bottom: 3-D representation of actual $E_k - \hat{E}_k$.

8.2 Iteration-varying Model Uncertainty

In this section, we consider the problem of monotonic convergence for ILC systems that face stochastic noise along the iteration axis. In most of the existing work in the literature the model uncertainty has been considered to be iteration-invariant. Likewise, in previous chapters, the MC property of the ILC process has been investigated based on the Lyapunov equation, vertex matrices of the interval Markov matrix, and norm-based methods. However, the scope of those works was restricted to iteration-invariant model uncertainties. Yet, it is quite natural to suppose that model uncertainty could be iteration-variant and there could be non-deterministic model variation at every iteration. This section is dedicated to this problem. In particular, we try to guarantee MC of ILC for iteration-varying model uncertain systems until the tracking error is reduced below a previously calculated baseline error boundary.

8.2.1 Basic Background Materials of Model Uncertain Super-vector ILC

Let us consider the following 2-dimensional uncertain system:

$$x_k(t+1) = (A + \Delta_k^A)x_k(t) + (B + \Delta_k^B)u_k(t) + v(k,t) \tag{8.42}$$
$$y_k(t) = (C + \Delta_k^C)x_k(t) + w(k,t) \tag{8.43}$$

where Δ_k^A, Δ_k^B, and Δ_k^C are uniformly distributed (stochastic) model uncertainties varying along the iteration axis, and $v(k,t)$ and $w(k,t)$ are time- and iteration-dependent process and measurement noises. As a reminder, t is the discrete-time point along the time axis, which means that t is on a finite horizon. That is, at each iteration trial, t has N different discrete-time points. Meanwhile, k is on an infinite horizon. So it monotonically increases. Thus, we can consider two different types of model uncertainties. The first type is iteration-independent model uncertainty and the second type is iteration-variant model uncertainty. In this section, we are interested in the latter. Under this research direction, the first required task is to convert the model uncertainty of (8.42)–(8.43) to the Markov matrix H_k according to model uncertainty type. Let us suppose that the model uncertainty has been incorporated into the system Markov matrix properly (see Chapter 4) and that we can write

$$H_k = H^o + \Delta_k^H, \tag{8.44}$$

where H^o is the system Markov matrix corresponding to the nominal plant; Δ_k^H is a uniformly distributed (stochastic) additive uncertain Markov matrix corresponding to the uncertainty of the plant, which is bounded as $\Delta_k^H \in \Delta^I := [\underline{\Delta}, \overline{\Delta}]$ with upper boundary $\overline{\Delta}$ and lower boundary $\underline{\Delta}$. Superscript I is used to denote the interval model uncertainty as in Chapter 4, but in this case the variation between the boundaries is stochastic (with respect to the iteration axis). Then, using a model uncertain set, H_k can be defined as

$$H_k \in H^I := [\underline{H}, \overline{H}], \tag{8.45}$$

where H^I is an interval uncertain Markov plant set, $\underline{H}$ is the lower boundary of the interval Markov plant, and $\overline{H}$ is the upper boundary of the interval Markov plant. Next, our task is to find the learning gain matrix such that the uncertain super-vector ILC system defined from (8.44) is asymptotically stable or MC against all possible uncertain interval plants H^I. Throughout this section, it is assumed that $\overline{\Delta} = -\underline{\Delta}$ and the model uncertainty is bounded in an operator norm topology (without special comment, $\| \cdot \|$ represents an operator 2-norm) such as

$$\|\Delta_k^H\| \le \Delta^*, \ \Delta_k^H \in \Delta^I, k = 1, 2, \ldots, \infty,$$

where Δ^* is the maximum singular value of the interval matrix Δ^I, which can be calculated from the following lemma:

Lemma 8.16. *The maximum 2-norm of Δ_k^H, i.e., Δ^*, is calculated from* $\|\underline{\Delta}\| = \|\overline{\Delta}\| = \Delta^*$. ∎

Proof. See Theorem 3.1 of [384]. ∎

8.2.2 ILC Design for Iteration-varying Model Uncertain System

In this section, iteration-varying learning gain matrices are designed. Note that, although it is not impossible to guarantee the MC of iteration-varying model uncertain ILC system with fixed learning gain matrices,[3] we have found that it is quite difficult to reduce the baseline error using fixed learning gain matrices. Thus, this section focuses on the iteration-varying learning gain matrices. In the sequel, the main theoretical developments are briefly summarized.

Let us first denote iteration-varying Markov plants by $H_k \equiv H^o + \Delta_k$. Then, using the ILC update rule

$$U_{k+1} = \Lambda_k U_k + \Gamma_k E_k, \tag{8.46}$$

we can derive the following error propagation rule:

$$E_{k+1} = \left(H_{k+1}\Lambda_k(H_k)^{-1} - H_{k+1}\Gamma_k\right) E_k + \left(I - H_{k+1}\Lambda_k(H_k)^{-1}\right) Y_d. \tag{8.47}$$

In (8.46), the ILC control input U_{k+1} is calculated by multiplying the control-signal proportional gain matrix Λ_k by U_k and by multiplying the error-signal proportional gain matrix Γ_k by E_k. Here, notice that desired trajectory Y_d is fixed and tracking error E_k is assumed known because we calculate the control signal of the $(k+1)^{\text{th}}$ iteration trial using the past information before the kth iteration trial. Now, since Y_d and U_k are known, if it is assumed that H_k can be estimated at the $(k+1)^{\text{th}}$ iteration, then we know that during the calculation of the propagated $(k+1)^{\text{th}}$ error signal in (8.47), only the Markov plant matrix H_{k+1} at the $(k+1)^{\text{th}}$ iteration is unknown. From this observation, we rewrite (8.47) as

$$E_{k+1} = \left(H_{k+1}\Lambda_k(\widehat{H}_k)^{-1} - H_{k+1}\Gamma_k\right) E_k + \left(I - H_{k+1}\Lambda_k(\widehat{H}_k)^{-1}\right) W_k E_k, \tag{8.48}$$

where $\widehat{H}_k$ is an estimate of the Markov plant H_k at the $(k+1)^{\text{th}}$ iteration. Then, introducing an intermediary matrix W_k that satisfies $Y_d = W_k E_k$, we obtain the following theorem:

---

[3] In the preceding section, we provided a scheme for reducing the baseline error with a fixed learning gain matrix, but with iteration-independent model uncertainty. That is, in the preceding section the iteration-varying uncertainty was due to disturbances and noise, not model variation. In this section the uncertainty is in the model.

Theorem 8.17. *If the learning gain matrices are given as* $\Lambda_k = (H^o)^{-1}\widehat{H}_k$ *and* $\Gamma_k = (H^o)^{-1}$, *then the iteration-varying ILC system is MC, i.e.,* $\|E_{k+1}\| < \|E_k\|$ *if*

$$\Delta^* < \frac{1}{\alpha}$$

where $\alpha = \min_{W_k}\{\|(H^o)^{-1}W_k\|\}$ *and* W_k *is determined from* $Y_d = W_k E_k$. ∎

Proof. From (8.48), the MC condition is given as

$$\|\mathcal{T}\| < 1, \tag{8.49}$$

where $\mathcal{T} := H_{k+1}\Lambda_k(\widehat{H}_k)^{-1} - H_{k+1}\Gamma_k + \left(I - H_{k+1}\Lambda_k(\widehat{H}_k)^{-1}\right)W_k$. Let us substitute $\Gamma_k = \Lambda_k\widehat{H}_k^{-1}$ into (8.49). Then we have

$$\left\|\left(I - H_{k+1}\Lambda_k(\widehat{H}_k)^{-1}\right)W_k\right\| < 1.$$

From the relationships:

$$
\begin{aligned}
\left\|\left(I - H_{k+1}\Lambda_k(\widehat{H}_k)^{-1}\right)W_k\right\| &= \left\|\left(I - (H^o + \Delta_{k+1})\Lambda_k(\widehat{H}_k)^{-1}\right)W_k\right\| \\
&= \left\|\left(I - (H^o + \Delta_{k+1})(H^o)^{-1}\right)W_k\right\| \\
&\quad - \left\|\Lambda_{k+1}(H^o)^{-1}W_k\right\| \\
&\leq \|\Delta_{k+1}\|\left\|(H^o)^{-1}W_k\right\| \\
&\leq \Delta^*\left\|(H^o)^{-1}W_k\right\|,
\end{aligned}
$$

if

$$\Delta^*\left\|(H^o)^{-1}W_k\right\| < 1 \Leftrightarrow \Delta^* < \frac{1}{\left\|(H^o)^{-1}W_k\right\|},$$

then the ILC process is MC. Thus, we have $\Gamma_k = (H^o)^{-1}$. However, observe that the W_k matrix that satisfies the relationship $Y_d = W_k E_k$ is not unique. Therefore, to check the MC of the ILC process at the kth iteration without conservatism, we have to find

$$\max_{W_k}\left\{\frac{1}{\|(H^o)^{-1}W_k\|}\right\} \Leftrightarrow \min_{W_k}\left\{\|(H^o)^{-1}W_k\|\right\},$$

with $Y_d = W_k E_k$. Consequently, at the kth iteration trial, if

$$\Delta^* < \frac{1}{\alpha}$$

where $\alpha = \min_{W_k}\{\|(H^o)^{-1}W_k\|\}$ with $Y_d = W_k E_k$, then $\|E_{k+1}\| < \|E_k\|$. ∎

If we can allow some conservatism, Theorem 8.17 can be relaxed. From $\|Y_d\| \leq \|W_k\|\|E_k\|$ and from

$$\frac{1}{\|(H^o)^{-1}\|\|W_k\|} \leq \frac{1}{\|(H^o)^{-1}W_k\|},$$

we derive the following relaxed result.

Corollary 8.18. *With $\Lambda_k = (H^o)^{-1}\widehat{H}_k$ and $\Gamma_k = (H^o)^{-1}$, the iteration-varying ILC system is MC if*

$$\Delta^* < \frac{\underline{\sigma}(H^o)}{\alpha}$$

where $\alpha = \min\{\|W_k\|\}$ with $Y_d = W_k E_k$. ∎

Proof. Since $\|(H^o)^{-1}\| = \frac{1}{\underline{\sigma}(H^o)}$, the proof is immediate. ∎

Remark 8.19. In Theorem 8.17 and Corollary 8.18, the MC conditions were derived without calculating the inverse interval matrix of (8.47). Thus, the results will be much less conservative than a result obtained by calculating the inverse interval matrix of (8.47). However, it is a disadvantage of Theorem 8.17 and Corollary 8.18 that we must solve $\alpha = \min_{W_k}\{\|(H^o)^{-1}W_k\|\}$ and $\alpha = \min\{\|W_k\|\}$ with the constraint $Y_d = W_k E_k$ at every iteration trial.

Theorem 8.17 and Corollary 8.18 provide MC conditions at the kth iteration. Using some operator norm properties, we can also answer the reverse question: under what condition is MC not guaranteed? The following corollary answers this question:

Corollary 8.20. *With $\Lambda_k = (H^o)^{-1}\widehat{H}_k$ and $\Gamma_k = (H^o)^{-1}$, defining $E_k = \widehat{E}_k \varrho$ such that $\|\widehat{E}_k\| = 1$, if*

$$\frac{\underline{\sigma}(H^o)}{\alpha_{min}} \le \Delta^*$$

where $\alpha_{min} = \frac{\|Y_d\|}{\varrho}$, then MC of the ILC process is not guaranteed. ∎

Proof. From Corollary 8.18, $\alpha = \min\{\|W_k\|\}$ with $Y_d = W_k E_k$. Now, changing this into $Y_d = W_k \bar{E}_k \varrho$ where $\|\bar{E}_k\| = 1$, we find the relationship

$$\|Y_d\| = \|W_k \bar{E}_k\|\varrho \le \|W_k\|\varrho, \tag{8.50}$$

because $\|W_k\| = \max_{\|x\|=1}\|W_k x\|$. Thus, $\alpha = \min\{\|W_k\|\} \ge \frac{\|Y_d\|}{\varrho}$ with any W_k if W_k satisfies the equality $Y_d = W_k E_k$, because $\|W_k\| \ge \frac{\|Y_d\|}{\varrho}$. Therefore, the following relationship is always true: $\frac{\underline{\sigma}(H^o)}{\alpha} \le \frac{\underline{\sigma}(H^o)}{\alpha_{min}}$. Consequently, if $\frac{\underline{\sigma}(H^o)}{\alpha_{min}} \le \Delta^*$, then $\frac{\underline{\sigma}(H^o)}{\alpha} \le \Delta^*$, so MC is not guaranteed. ∎

In Theorem 8.17 and Corollary 8.18, to calculate α, an optimization scheme can be used. However, in each iteration, it is computationally expensive to perform the optimization. In what follows, a sub-optimal scheme is suggested to find α. To find $\alpha = \min_{W_k}\{\|(H^o)^{-1}W_k\|\}$, with the constraint $Y_d = W_k E_k$, let us use the equality:

$$(H^o)^{-1}Y_d = (H^o)^{-1}W_k E_k. \tag{8.51}$$

Then the sub-optimal strategy uses the inequality $\|(H^o)^{-1}W_k\| \le \|(H^o)^{-1}\| \|W_k\| \le \|(H^o)^{-1}\| \sqrt{N}\|W_k\|_\infty$, where N is the size of the matrix W_k and $\|\cdot\|_\infty$ is the operator ∞-norm. We will minimize $\|(H^o)^{-1}W_k\|$ by minimizing $\|W_k\|_\infty$ because $\|(H^o)^{-1}\|$ is fixed.

Theorem 8.21. *Let us define the maximum absolute element of E_k as*

$$\beta \equiv \arg_{E_k(j)} \max_{i=1,\ldots,N} |E_k(i)|,$$

where $E_k = [E_k(1), E_k(2), \ldots, E_k(N)]^T$ and $j \in \{1, 2, \ldots, N\}$. Also define an $N \times N$ matrix $\mathcal{O}$, whose elements are zero except its jth column (denoted by $\mathcal{O}^j$), by

$$\mathcal{O}^j \equiv [Z(1)/\beta, Z(2)/\beta, \ldots, Z(N)/\beta]^T, \qquad (8.52)$$

where the vector Z is given as $Z \equiv (H^o)^{-1}Y_d$. Then, a sub-optimal α for Theorem 8.17 and Corollary 8.18 is calculated as $\alpha = \|\mathcal{O}\|$. ■

Proof. In Theorem 8.17, we are interested in finding $\min_{W_k} \{\|(H^o)^{-1}W_k\|\}$, subject to $(H^o)^{-1}Y_d = (H^o)^{-1}W_k E_k$. In $Y_d = W_k E_k$, Y_d, and E_k are known vectors. Observe that, as already commented, the W_k that satisfies $Y_d = W_k E_k$ is not unique. Let us consider the first row vector of W_k. It is easy to see that a vector $W_k(1, i), i = 1, \ldots, N$ that satisfies

$$Y_d(1) = W_k(1, 1)E_k(1) + W_k(1, 2)E_k(2) + \cdots + W_k(1, N)E_k(N),$$

where $Y_d(1)$ is the first element of vector Y_d and $E_k(i)$ are elements of E_k, can be solutions for the first row vector (denoted W_k^1) of W_k. We can arbitrarily define $W_k(1, i)$ as $W_k(1, 1) = \frac{\zeta_1 Y_d(1)}{E_k(1)}$, $W_k(1, 2) = \frac{\zeta_2 Y_d(1)}{E_k(2)}$, ..., $W_k(1, N) = \frac{\zeta_N Y_d(1)}{E_k(N)}$, where $\sum_{i=1}^{N} \zeta_i = 1$. In this case, the 1-norm of the first row vector of W_k is calculated as

$$\|W_k^1\|_1 = \sum_{i=1}^{N} |W_k(1, i)| = \sum_{i=1}^{N} \left| \frac{\zeta_i Y_d(1)}{E_k(i)} \right|. \qquad (8.53)$$

Here, the following relationship is always true regardless of ζ_i:

$$\sum_{i=1}^{N} \left| \frac{\zeta_i Y_d(1)}{E_k(i)} \right| \geq \sum_{i=1}^{N} \left| \frac{\zeta_i Y_d(1)}{\beta} \right| \geq \left| \frac{Y_d(1)}{\beta} \right|. \qquad (8.54)$$

Hence, from (8.53) and (8.54), we choose $\min\{\|W_k^1\|_1\}$ as $\min\{\|W_k^1\|_1\} = \left| \frac{Y_d(1)}{\beta} \right|$. Therefore, if index j determined from $\beta \equiv \arg_{E_k(j)} \max_{i=1,\ldots,N} |E_k(i)|$ is found, the first row vector of W_k can be determined as $W_k(1, 1) = 0, \ldots, W_k(1, j-1) = 0, W_k(1, j) = 1, W_k(1, j+1) = 0, \ldots, W_k(1, N) = 0$. In the same way, the generalized minimum of the pth row vector (denoted as W_k^p) of W_k is also determined as $W_k(p, 1) = 0, \ldots, W_k(p, j-1) = 0, W_k(p, j) = 1, W_k(p, j+1) = 0, \ldots, W_k(p, N) = 0$. Finally, since $\|W_k\|_\infty = \max_{1 \leq i \leq N} \sum_{j=1}^{N} |W_k(i, j)| = \max_{1 \leq i \leq N} \|W_k^i\|_1$, from (8.54), we can find $\min\{\|W_k\|_\infty\}$ when the column vectors of W_k are zero except for the jth column vector, which is calculated as $[Y_d(1)/\beta, Y_d(2)/\beta, \ldots, Y_d(N)/\beta]^T$. Therefore, $(H^o)^{-1}W_k = \mathcal{O}$. This completes the proof. ■

Now, by simple algebraic manipulation, since the following is true:

$$\|\mathcal{O}\| = \frac{1}{|\beta|}\|Z\|$$

we can obtain a simplified MC condition:

Corollary 8.22. *If* $\Delta^* \leq \frac{|\beta|}{\|Z\|}$, *the learning gain matrices* $\Lambda_k = (H^o)^{-1}\widehat{H}_k$ *and* $\Gamma_k = (H^o)^{-1}$ *guarantee MC of the iteration-varying ILC system.* ∎

Corollary 8.22 provides a very simple MC checking method. Now, based on the discussion above, we arrive at our final result for the simplified MC condition and a baseline error boundary.

Theorem 8.23. *Learning gain matrices* $\Lambda_k = (H^o)^{-1}\widehat{H}_k$ *and* $\Gamma_k = (H^o)^{-1}$ *guarantee MC if*

$$\Delta^*\|Z\| < \|E_k\|_\infty,$$

and the baseline error boundary E^* *is then calculated as* $E^* = \Delta^*\|Z\|$. ∎

Proof. From $\Delta^* \leq \frac{|\beta|}{\|Z\|}$, since $\Delta^*\|Z\| \leq |\beta|$ and $|\beta| = \|E_k\|_\infty$ are true, $\Delta^*\|Z\| < \|E_k\|_\infty$ is straightforward. Also, we know that if $\|E_k\|_\infty$ is bigger than $\Delta^*\|Z\|$, which can be calculated previously, the ILC process is then MC. Hence, the system converges until the tracking error is reduced below $\Delta^*\|Z\|$. Thus, the baseline error boundary in the ∞-norm topology can be calculated as $E^* = \Delta^*\|Z\|$. ∎

8.2.3 Parameter Estimation

In this section, a method is developed to estimate the Markov parameters of the plant to find $\widehat{H}_k$. Let us consider the following stochastic system:

$$Y_k = H_k U_k + w_k \tag{8.55}$$

where w_k is the measurement noise. Here, we seek an optimal estimation of the Markov matrix H_k using the known control input U_k and the measured output Y_k, but with stochastic noise w_k. In this monograph, for a simple result, we use the Wiener filter and the least squares method. Generally, in the Wiener filter or least squares estimation, if the system is given as

$$u = Hv + w \tag{8.56}$$

where u is the output measurement, H is the system, v is the input, and w is stochastic noise, then it is assumed that u is measured and H is known. In this case, the Wiener filter or least squares method can be used to estimate the input signal v effectively. However, in our problem, it is assumed that H is unknown, but u is measured and v is known. Thus, clearly the problem setup of our case is different from the general Wiener filter or least squares method. However, fortunately, due to the fact that H_k is a lower triangular Toeplitz matrix, we can easily formulate the Wiener filter for our ILC system. The key idea is to use the following property:

Property 8.24. If $U_k(1) \neq 0$, then the following commutative property is true:

$$H_k U_k = \mathbf{U}_k \mathbf{h}_k \tag{8.57}$$

where H_k is a lower triangular Toeplitz matrix, U_k is a column vector, $\mathbf{U}_k$ is a Toeplitz matrix defined from U_k by:

$$\mathbf{U}_k = \begin{bmatrix} U_k(1) & 0 & 0 & \cdots & 0 \\ U_k(2) & U_k(1) & 0 & \cdots & 0 \\ U_k(3) & U_k(2) & U_k(1) & \cdots & 0 \\ \vdots & \vdots & \vdots & \ddots & \vdots \\ U_k(N) & U_k(N-1) & U_k(N-2) & \cdots & U_k(1) \end{bmatrix}$$

and the new Markov column vector $\mathbf{h}_k$ is defined as

$$\mathbf{h}_k = [H_k(1,1), H_k(2,1), H_k(3,1), \dots, H_k(N,1)]^T. \tag{8.58}$$

Now, using Property 8.24, (8.55) can be rewritten as

$$Y_k = H_k U_k + w_k = \mathbf{U}_k \mathbf{h}_k + w_k. \tag{8.59}$$

Finally, by estimating $\mathbf{h}_k$ from the Wiener filter or the least squares method (we do not explain the Wiener filter and least squares methods in detail; for these methods, for example, see [214]), we can optimally estimate the Markov matrix H_k for the $(k+1)^{\text{th}}$ iteration trial. From the least squares method, $\mathbf{h}_k$ is estimated as

$$\widehat{\mathbf{h}}_k = (\mathbf{U}_k^T \mathbf{U}_k)^{-1} \mathbf{U}_k^T \widetilde{Y}_k. \tag{8.60}$$

Using the Wiener filter, the Markov parameters are estimated as

$$\widehat{\mathbf{h}}_k = \mathcal{G} \widetilde{Y}_k, \tag{8.61}$$

where the Wiener filter $\mathcal{G}$ is calculated by

$$\mathcal{G} = [\mathbf{U}_k^T \mathbf{U}_k + \sigma_n^2 \mathbf{R}^{-1}]^{-1} \mathbf{U}_k^T, \tag{8.62}$$

with

$$\mathbf{R} = E[\mathbf{h}_k \mathbf{h}_k^T], \quad \mathbf{R}_n = \text{diag}(\sigma_n^2) = E[w_k w_k^T].$$

Observe that if the least squares method is used, we do not need to know the stochastic characteristics of the uncertain Markov matrix and measurement noise, while when using the Wiener filter, we need to know these stochastic characteristics. It is also necessary to pay attention to the calculation of $E[\mathbf{h}_k \mathbf{h}_k^T]$ because $\mathbf{h}_k$ is an interval vector defined as $\mathbf{h}_k = \mathbf{h}_k^o + \Delta \mathbf{h}_k$ or $\mathbf{h}_k = [\underline{\mathbf{h}}_k, \overline{\mathbf{h}}_k]$. Here, simply by assuming uniformly distributed interval uncertainty, we calculate the expectation value such $E[\mathbf{h}_k] = \mathbf{h}_k^o$. Thus, $\mathbf{R}$ is simply calculated as

$$\mathbf{R} = E[\mathbf{h}_k \mathbf{h}_k^T] = \begin{bmatrix} h_1^o h_1^o & h_1^o h_2^o & \cdots & h_1^o h_N^o \\ h_2^o h_1^o & h_2^o h_2^o & \cdots & h_2^o h_N^o \\ \vdots & \vdots & \ddots & \vdots \\ h_N^o h_1^o & h_N^o h_2^o & \cdots & h_N^o h_N^o \end{bmatrix}. \tag{8.63}$$

Remark 8.25. In the Wiener filter, for the calculation of $\mathcal{G}$ we need to find the inverse of $\mathbf{R}$. But, as shown in (8.63), rank$(\mathbf{R}) = 1$, so the inverse of $\mathbf{R}$ does not exist. To overcome this singularity, we give a bit of interval uncertainty to each element of $\mathbf{R}$.

Remark 8.26. In the parameter estimation, it is assumed that $U_k(1) \neq 0$ in order to avoid the singularity of $\mathbf{U}_k$. Thus, during initial iteration, the first discrete-time control input is forced to satisfy $U_1(1) \neq 0$, while $U_1(i) = 0$ is allowed when $i = 2, 3, \ldots, N$.

8.2.4 Simulation Illustrations

We use the same example given in Section 5.3. To avoid the singularity mentioned in Remark 8.26, the initial control input at the first iteration is given as 1, i.e., $U_1(1) = 1$, and $U_1(i) = 0$, $i = 2, 3, \ldots, N$ and the measured noise covariance is modeled as $\sigma_n^2 = 0.001$. From (8.40) and (8.41), the nominal Markov parameters are calculated as $\mathbf{h}^o = [2.0000, 1.6092, -0.0235, -1.0295, -1.3839, -0.8676, -0.0421, 0.6322, 0.8118, 0.5263]^T$. For the interval model uncertainty, we allowed 10 percent interval uncertainty at each iteration. For the interval uncertainty, the MATLAB® command *rand* was used and for the measured noise, the MATLAB® command *randn* was used. Figure 8.5 shows the test result. In this figure, $\Delta^* \|Z\|$ is calculated from Theorem 8.23. From this result, we find that the suggested learning gain matrices and parameter estimation method guarantee the baseline error to be below the previously calculated value. Further, until the baseline error is reduced below the calculated baseline boundary, the tracking error is converging monotonically.

8.2.5 Concluding Remarks

In this section, MC ILC algorithms were designed for plants with iteration-varying model uncertainty. Using the suggested method, we calculated the baseline error boundary and showed that the tracking error is reduced below this baseline. Moreover, the MC property of the ILC process is enforced until it arrives the baseline error boundary. Thus, although there exists iteration-varying model uncertainty, the ILC approach suggested in this section provides a satisfactory transient performance.

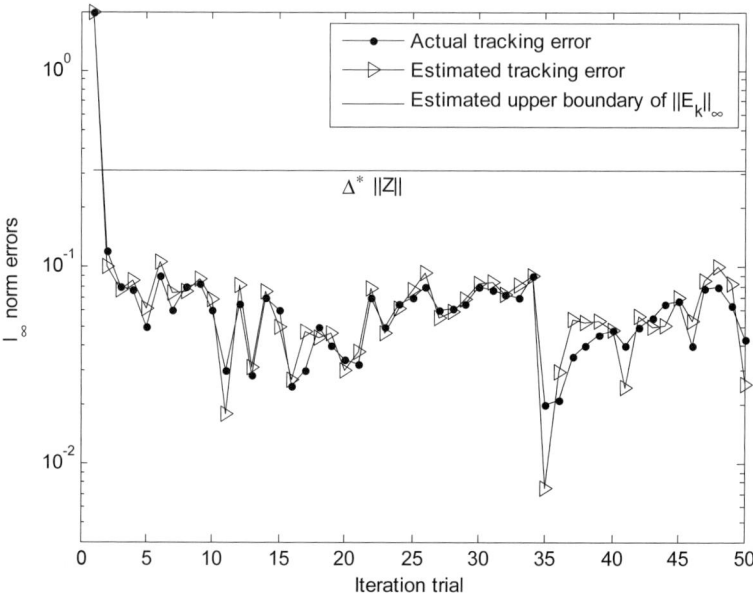

Fig. 8.5. Baseline error of iteration-varying model uncertain ILC system

8.3 Intermittent Iterative Learning Control

Robustness has been studied in ILC from a number of different perspectives, depending on the underlying assumptions made about the system to be controlled. For example, in stochastic ILC, learning gain matrices are designed using Kalman filters to primarily take into account measurement noise [397, 396, 395, 54]. In H_∞ ILC, external disturbances and model uncertainty are considered on the time axis [360]. Robustness with respect to initial resetting has been one of the most important research topics in ILC, given that by assumption ILC requires an initial equivalent condition at every iteration [353, 75, 131, 431, 434]. Frequency-domain analysis and/or synthesis based on frequency-based filtering considers noise and disturbances in the frequency domain [377, 127, 327, 330].

In this section, we introduce a new type of uncertainty that can arise during the implementation of an ILC system – data dropout during transmission of measurements or control signals between a remote plant and the ILC controller – and we provide an analytical ILC learning gain design method to ensure convergence in the face of such uncertainty. The section is organized as follows: In Section 8.3.1, we explain the idea of intermittent ILC in detail, including the motivation for this study. In Section 8.3.2, we provide a synthesis method and an analysis for convergence. Using a stochastic Kalman filtering approach, we show the surprising result that it is possible to design a learning gain such that the system eventually converges to a desired trajectory as long as all the data is not lost. Conclusions are given in Section 8.3.3.

8.3.1 Intermittent ILC

In this subsection, we provide motivations for the problem of intermittent ILC.

Networked Control Systems

Recently, networked control systems (NCS) have become very popular, due to the benefit of reducing the complexity of directly wiring between computers, saving the costs of maintenance of controllers at remote plants, etc. Due to these benefits, real-time industrial networks such as DeviceNet, Profibus, FireWire, etc., have emerged as new technologies for distributed control applications [46]. Most of these industrial networks have been used for remote control applications and factory automation. The key feature of these industrial networks is to connect sensors, actuators, and controllers as network-wired nodes. This feature has enabled reducing the system wiring, increasing system agility, making it easier to diagnose the system, and increasing system reliability. Applications of computer networks include military unmanned vehicles, manufacturing plants, telerobotics, telemedicine, and various kinds of information and data signal exchanges between spatially distributed system components. Generally, the distributed system consists of a supervisory controller, a remote plant, the main controller, actuators, sensors, and a network that connects all these components. Figure 8.6 depicts this idea.

Unfortunately, despite the benefits, in NCS applications there are also some drawbacks. First, there is the problem of data congestion, which is caused by lack of a universal clock between the main controller and the remote plant, hardware-inherent data delays, and communication constraints such as channel capacity. Furthermore, in the case of MIMO systems or multiple networked plants, data congestion and other factors, such as hard-nonlinearities, can cause time-asynchrony among the multiple remote plants or subsystems. One of the major effects of these drawbacks is data dropout. In NCS research there have been numerous efforts to compensate for data congestion [447, 456, 491, 46, 470, 283, 471, 231, 339], and to improve the performance of systems

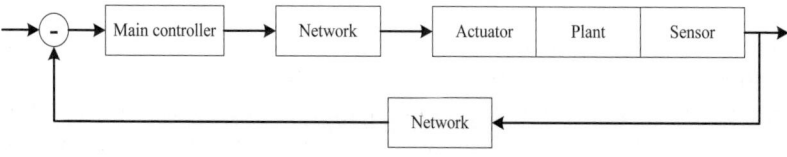

Fig. 8.6. Direct closed-loop networked control system configuration

that experience data dropout [463, 281, 273, 272, 271, 552, 411, 482, 412]. With regard to data dropout, it has been shown that there is a critical data dropout rate [411, 482, 412] above which the networked system becomes unstable and therefore the desired performance is no longer achieved.

In this section, we suggest that for iterative learning control systems operated over a networked control system it is possible to address, and, indeed, to alleviate the data dropout problem. Specifically, we show that if the plant in a networked control system is to be operated repetitively, then there is an incredible advantage to using ILC. Our main theoretical result is that if the control gain is updated by an ILC scheme and the remote plant operates in a repetitive way, the resulting NCS will be stable as long as there is a part of the data measurement, whatever amount it is, that gets through. That is, in an extreme case, even though 99 percent of the data is dropped, if there is 1 percent data measurement, the ILC networked control system will be stable along the iteration axis. This result is surprising and cannot be true in non-repetitive NCS problems [411, 482, 412].

What Is Intermittent ILC?

The key idea of ILC is to use information from past repetitions to compute the current repetition's control effort. Throughout this subsection, let us consider the following discrete-time, 2-dimensional system:

$$x_k(t+1) = Ax_k(t) + Bu_k(t) \tag{8.64}$$

$$y_k(t) = Cx_k(t) \tag{8.65}$$

where $A \in \mathcal{R}^{n \times n}, B \in \mathcal{R}^{n \times m}$, and $C \in \mathcal{R}^{l \times n}$. We can consider two different types of data dropouts when the ILC main controller is connected to the remote plant via a network. The first one occurs when the control input $u_k(t)$ is updated. For instance, in a typical ILC update law, the control input $u_{k+1}(t)$ is calculated by $u_{k+1}(t) = u_k(t) + \gamma e_k(t+1)$, where $e_k(t+1) = y_d(t+1) - y_k(t+1)$, y_d is the desired output, and y_k is the actual measured output. In this update, the stored control input information $u_k(t)$ or the error term $\gamma e_k(t+1)$ may be missed during the data transfer in the network. From Figure 8.6, the remote ILC controller is located in the box named "Main controller" and the controlled-remote plant is located in the box named "Plant." During the control signal transmission through the network, the signal could be dropped, which means that in the ILC control signal calculated by $u_{k+1} = u_k(t) + \gamma e_k(t+1)$, the past control signal $u_k(t)$ and/or error term $\gamma e_k(t+1)$ may be missed. The second type of data dropout is due to measurement data loss during the signal transfer from the sensor to the main controller. In this case, mathematically, equation (8.65) can be changed as

$$y_k(t) = \eta(t)Cx_k(t) \tag{8.66}$$

where $\eta(t)$ could be zero or one (i.e., $\eta(t) \in \{0,1\}$). But, this zero or one could be random due to the stochastic characteristic of the data loss. Thus, when we say *intermittent ILC*, we have to consider the above two different data dropout situations. However, in this monograph, for simplicity, we consider only the second case (intermittent measurement), as was also done in existing NCS works [463, 281, 273, 272, 271, 552, 411, 482, 412]. Note that if we use a higher-order ILC updating scheme, we may have to consider intermittent data dropouts from all past control signals and/or all past output measurements, but this is beyond the scope of this monograph.

8.3.2 Optimal Learning Gain Matrix Design for Intermittent ILC

For the intermittent ILC design, we use an existing Kalman filtering ILC approach. Our main theoretical contribution is to add the intermittent measurement signal to the ILC update and then analyze the overall convergence property of the designed intermittent ILC system. The result of this section is influenced by [395]. Specifically, we will design the ILC learning gain $K_k(t)$ with stochastic noises and intermittent measurements.[4]

Design of an Optimal Learning Gain

Throughout this subsection, we use the following update rule:

$$u_{k+1}(t) = u_k(t) + K_k(t)\eta e_k(t+1) \qquad (8.67)$$

where $\eta \in \{0,1\}$. Following standard ILC practice, use $u_d(t)$, $x_d(t)$, and $y_d(t)$ to denote the desired input, state, and output signals, respectively. Introducing $\delta u_{k+1}(t) = u_d(t) - u_{k+1}(t)$ and $\delta x_k(t) = x_d(t) - x_k(t)$, and following [395], we obtain the auxiliary system:

$$\begin{bmatrix} \delta u_{k+1}(t) \\ \delta x_k(t+1) \end{bmatrix} = \begin{bmatrix} I - K_k(t)\eta CB & -K_k(t)\eta CA \\ B & A \end{bmatrix} \times \begin{bmatrix} \delta u_k(t) \\ \delta x_k(t) \end{bmatrix}$$
$$+ \begin{bmatrix} K_k(t)\eta C & K_k(t) \\ -I & 0 \end{bmatrix} \times \begin{bmatrix} w_k(t) \\ v_k(t+1) \end{bmatrix}.$$

Now, we introduce $\eta = \bar{\eta} + \tilde{\eta}$, where $\bar{\eta}$ is the mean of η and $\tilde{\eta}$ is a zero-mean stochastic sequence. Following [482], we can calculate the variance of $\tilde{\eta}$ as $\sigma_{\tilde{\eta}}^2 = (1 - \bar{\eta})\bar{\eta}$. Now, inserting $\eta = \bar{\eta} + \tilde{\eta}$ into (8.68) yields

---

[4] In [395], there is no mathematical derivation considering intermittent measurements. Thus, the contribution of this section over [395] is to introduce intermittent measurements into the stochastic ILC framework. Also we note again, the contribution of this monograph over existing intermittent Kalman filtering [463, 281, 273, 272, 271, 552, 411, 482, 412] is to show that there is no critical data dropout rate in the point of stability when we enhance the intermittent Kalman filtering results with an ILC update scheme.

$$\begin{bmatrix} \delta u_{k+1}(t) \\ \delta x_k(t+1) \end{bmatrix} = \begin{bmatrix} I - K_k(t)\overline{\eta}CB & -K_k(t)\overline{\eta}CA \\ B & A \end{bmatrix} \begin{bmatrix} \delta u_k(t) \\ \delta x_k(t) \end{bmatrix}$$

$$+ \begin{bmatrix} K_k(t)\overline{\eta}C & K_k(t) \\ -I & 0 \end{bmatrix} \begin{bmatrix} w_k(t) \\ v_k(t+1) \end{bmatrix}$$

$$+ \begin{bmatrix} -K_k(t)\widetilde{\eta}CB & -K_k(t)\widetilde{\eta}CA \\ 0 & 0 \end{bmatrix} \begin{bmatrix} \delta u_k(t) \\ \delta x_k(t) \end{bmatrix}$$

$$+ \begin{bmatrix} K_k(t)\widetilde{\eta}C & 0 \\ 0 & 0 \end{bmatrix} \begin{bmatrix} w_k(t) \\ v_k(t+1) \end{bmatrix}. \tag{8.68}$$

For convenience, defining

$$\Phi := \begin{bmatrix} I - K_k(t)\overline{\eta}CB & -K_k(t)\overline{\eta}CA \\ B & A \end{bmatrix}$$

$$\Psi := \begin{bmatrix} K_k(t)\overline{\eta}C & K_k(t) \\ -I & 0 \end{bmatrix}$$

$$\Omega := \begin{bmatrix} -K_k(t)CB & -K_k(t)CA \\ 0 & 0 \end{bmatrix}$$

$$\Upsilon :- \begin{bmatrix} K_k(t)C & 0 \\ 0 & 0 \end{bmatrix}; \quad X^+ := \begin{bmatrix} \delta u_{k+1}(t) \\ \delta x_k(l+1) \end{bmatrix}$$

$$X := \begin{bmatrix} \delta u_k(t) \\ \delta x_k(t) \end{bmatrix}; \quad W := \begin{bmatrix} w_k(t) \\ v_k(t+1) \end{bmatrix},$$

we rewrite (8.68) as

$$X^+ = \Phi X + \widetilde{\eta}\Omega X + \Psi W + \widetilde{\eta}\Upsilon W. \tag{8.69}$$

Now, taking expectations on both sides of (8.69), and assuming no correlation among state, random sequence, and random noises, we have

$$E[X^+ X^{+T}] = E[(\Phi X + \widetilde{\eta}\Omega X + \Psi W + \widetilde{\eta}\Upsilon W)$$
$$\times (\Phi X + \widetilde{\eta}\Omega X + \Psi W + \widetilde{\eta}\Upsilon W)^T]$$
$$= \Phi E[XX^T]\Phi^T + \sigma_{\widetilde{\eta}}^2 \Omega E[XX^T]\Omega^T$$
$$+ \Psi E[WW^T]\Psi^T + \sigma_{\widetilde{\eta}}^2 \Upsilon E[WW^T]\Upsilon^T. \tag{8.70}$$

Assuming that we have knowledge of the noise statistics and denoting them as

$$P^+ := E[X^+ X^{+T}]; \quad P := E[XX^T]; \quad Q := E[WW^T],$$

we have

$$P^+ = \Phi P \Phi^T + \sigma_{\widetilde{\eta}}^2 \Omega P \Omega^T + \Psi Q \Psi^T + \sigma_{\widetilde{\eta}}^2 \Upsilon Q \Upsilon^T. \tag{8.71}$$

To find an optimal learning gain matrix $K_k(t)$, following [395], we use the trace of P^+, which is the typical process in Kalman filtering. In what follows, for simplicity, we omit the subscripts k and $\tilde{\eta}$, and the time index t. Now, we compute the trace of both sides of (8.71) as:

$$\text{trace}(P^+) = \text{trace}[\Phi P \Phi^T + \sigma_{\tilde{\eta}}^2 \Omega P \Omega^T + \Psi Q \Psi^T + \sigma_{\tilde{\eta}}^2 \Upsilon Q \Upsilon^T].$$

Partitioning P and Q according to:

$$P = \begin{bmatrix} P_{11} & P_{12} \\ P_{21} & P_{22} \end{bmatrix}; \; Q = \begin{bmatrix} Q_{11} & Q_{12} \\ Q_{21} & Q_{22} \end{bmatrix}$$

we have:

$$
\begin{aligned}
\text{trace}(P^+) = \text{trace} \Big\{ & \sigma^2 KC(BP_{11} + AP_{21})B^T C^T K^T \\
& + \sigma^2 KC(BP_{12} + AP_{22})A^T C^T K^T \\
& + [(I - K\tilde{\eta}CB)P_{11} - K\tilde{\eta}CAP_{21}][I - K\tilde{\eta}CB]^T \\
& + [(I - K\tilde{\eta}CB)P_{12} - K\tilde{\eta}CAP_{22}][-K\tilde{\eta}CA]^T \\
& + (BP_{11} + AP_{21})B^T + (BP_{12} + AP_{22})A^T \\
& + (K\tilde{\eta}CQ_{11} + KQ_{21})(K\tilde{\eta}C)^T \\
& + (K\tilde{\eta}CQ_{12} + KQ_{22})K^T + Q_{11} \\
& + \sigma^2 KCQ_{11}(KC)^T \Big\}.
\end{aligned}
\tag{8.72}
$$

Next, substituting the following into the above equation:

$$V_1 := (CB, \; CA), \; V_2 := (B, \; A), \; V_3 = (I, \; 0),$$

we can simplify the right-hand side of (8.72) to be:

$$
\begin{aligned}
\text{trace} \Big\{ & \sigma^2 KV_1 P V_1^T K^T + \overline{\eta}^2 KV_1 P V_1^T K^T + V_2 P V_2^T \\
& + V_3 P V_3^T - \overline{\eta} V_3 P V_1^T K^T - \overline{\eta} K V_1 P V_3^T \\
& + Q_{11} + K(\overline{\eta}^2 C Q_{11} C^T + \sigma^2 C Q_{11} C^T + \overline{\eta} C Q_{12} \\
& + Q_{21} \overline{\eta} C^T + Q_{22})K^T \Big\}
\end{aligned}
\tag{8.73}
$$

Therefore, we have

$$
\begin{aligned}
\frac{\partial \text{trace}(P^+)}{\partial K} = & \; 2\sigma^2 KV_1 P V_1^T + 2\overline{\eta} K V_1 P V_1^T - 2\overline{\eta} V_3 P V_1^T \\
& + 2K(\overline{\eta}^2 C Q_{11} C^T + \sigma^2 C Q_{11} C^T + \overline{\eta} C Q_{12} \\
& + Q_{21} \overline{\eta} C^T + Q_{22}).
\end{aligned}
\tag{8.74}
$$

Here, assuming no correlation between $w_k(t)$ and $v_k(t)$, we consider $Q_{12} = Q_{21} = 0$. Now, defining

$$\Pi := (\sigma^2 + \overline{\eta}^2)V_1 P V_1^T + (\overline{\eta}^2 + \sigma^2)C Q_{11} C^T + Q_{22},$$

when $\frac{\partial \text{trace}(P^+)}{\partial K}$ is set equal to zero, we finally calculate the optimal learning gain as follows:

$$K_k(t) = \overline{\eta} V_3 P V_1^T \Pi^{-1}.
\tag{8.75}$$

Analysis of Convergence

In this subsection, we analyze the convergence of the intermittent ILC system updated by the learning gain given by (8.75). For this purpose, we need to find a recursive update formula for $P_{11} = E[\delta u_k(t)\delta u_k(t)^T]$. In fact, it is easy to show that $P_{12} = E[\delta u_k(t)\delta x_k(t)^T] = 0$, which can be proved using the same method given in [395]. Then, we obtain the following expansion:

$$\begin{aligned}
E[\delta u_{k+1}(t)\delta u_{k+1}(t)^T] = {} & \sigma^2 K V_1 P V_1^T K^T + \overline{\eta}^2 K V_1 P V_1^T K^T + V_3 P V_3^T \\
& - \overline{\eta} V_3 P V_1^T K^T - \overline{\eta} K V_1 P V_3^T + K(\overline{\eta}^2 C Q_{11} C^T \\
& + \sigma^2 C Q_{11} C^T + Q_{22}) K^T.
\end{aligned} \tag{8.76}$$

Now, reformulating the right-hand side of (8.76) and inserting (8.75) into this yields

$$E[\delta u_{k+1}(t)\delta u_{k+1}(t)^T] = V_3 P V_3^T - \overline{\eta}^2 V_3 P V_1^T \Pi^{-1} V_1 P V_3^T. \tag{8.77}$$

Further, writing $E[\delta u_k(t)\delta u_k(t)^T] = P_{11,k}$ and $E[\delta u_{k+1}(t)\delta u_{k+1}(t)^T] = P_{11,k+1}$, we have

$$P_{11,k+1} = (I - \overline{\eta} K_k(t) C B) P_{11,k}, \tag{8.78}$$

which is very similar to equation (16) of [395]. Here, writing $\Lambda := C Q_{11} C^T + (\sigma^2 + \overline{\eta}^2)^{-1} Q_{22}$ and $S := C A P_{22}(C A)^T + \Lambda$, we have

$$\begin{aligned}
I - \overline{\eta} K_k(t) C B = {} & I - \overline{\eta}^2 P_{11}(C B)^T (\sigma^2 + \overline{\eta}^2)^{-1} \\
& \times \left[C B P_{11}(C B)^T + S \right]^{-1} C B.
\end{aligned} \tag{8.79}$$

Now, since $\overline{\eta}^2/(\sigma^2 + \overline{\eta}^2) = \overline{\eta}$ and, for convenience, substituting $C B$ by N, then, the right-hand side becomes:

$$I - \overline{\eta} K_k(t) C B = I - \overline{\eta} P_{11} N^T \left[N P_{11} N^T + S \right]^{-1} N. \tag{8.80}$$

Next, using $\left[N P_{11} N^T + S \right]^{-1} = S^{-1} - S^{-1} N (N^T S^{-1} N + P_{11}^{-1})^{-1} N^T S^{-1}$, we obtain

$$\begin{aligned}
I - \overline{\eta} K_k(t) C B = {} & I - \overline{\eta} P_{11} N^T S^{-1} N + \overline{\eta} P_{11} N^T S^{-1} N \\
& \times (N^T S^{-1} N + P_{11}^{-1})^{-1} N^T S^{-1} N.
\end{aligned} \tag{8.81}$$

Also for convenience, writing $Y := N^T S^{-1} N$ and $W := P_{11} Y$, we finally obtain

$$\begin{aligned}
I - \overline{\eta} K_k(t) C B & = I - \overline{\eta} P_{11} Y + \overline{\eta} P_{11} Y (Y + P_{11}^{-1})^{-1} Y \\
& = I - \overline{\eta} P_{11} Y + \overline{\eta} P_{11} Y (I + (P_{11} Y)^{-1})^{-1} \\
& = I - \overline{\eta} W + \overline{\eta} W (I + W^{-1})^{-1} \\
& = (I - \overline{\eta} W)(I + W^{-1})(I + W^{-1})^{-1} + \overline{\eta} W (I + W^{-1})^{-1} \\
& = ((1 - \overline{\eta})I + W^{-1})(I + W^{-1})^{-1}.
\end{aligned} \tag{8.82}$$

Now it is observed that S is a positive definite matrix. Hence, if CB is full rank, then Y is positive definite (see page 399 of [200]). Also, it is assumed that the covariance matrix P_{11} is positive definite. Generally, the multiplication of two symmetric matrices is not symmetric. So, we cannot claim that W is symmetric. However, for SISO systems, W is a scalar. Then, we can conclude that $((1-\overline{\eta})I+W^{-1})(I+W^{-1})^{-1} < 1$; so $I-\overline{\eta}K_k(t)CB < 1$. Thus, $P_{11,k} \to 0$ in the manner of $P_{11,k+1} < P_{11,k}$ as $k \to \infty$. For the MIMO case, we prove $\rho\left[((1-\overline{\eta})I + W^{-1})(I + W^{-1})^{-1}\right] < 1$ in the following lemma.

Lemma 8.27. *The spectral radius of* $((1-\overline{\eta})I + W^{-1})(I + W^{-1})^{-1}$ *is less than 1.* ∎

Proof. Clearly, W^{-1} has positive real eigenvalues since it is multiplied by two positive definite matrices. Hence $(1-\overline{\eta})I+W^{-1}$ and $I+W^{-1}$ are nonsingular matrices. Let us say that X is an eigenvector matrix of $(1-\overline{\eta})I + W^{-1}$ and $\Sigma = \mathrm{diag}(\sigma_i)$ is diagonal eigenvalue matrix. Then, we can write

$$X[(1-\overline{\eta})I + W^{-1}]X^{-1} = \Sigma$$
$$\Leftrightarrow -\overline{\eta}XX^{-1} + X(I+W^{-1})X^{-1} = \Sigma$$
$$\Leftrightarrow X(I+W^{-1})X^{-1} = \mathrm{diag}(\overline{\eta} + \sigma_i). \tag{8.83}$$

Therefore, we have

$$(1-\overline{\eta})I + W^{-1} = X^{-1}\mathrm{diag}(\sigma_i)X \tag{8.84}$$

and

$$(I + W^{-1}) = X^{-1}\mathrm{diag}(\overline{\eta} + \sigma_i)X \Rightarrow (I + W^{-1})^{-1}$$
$$= X^{-1}\mathrm{diag}(\overline{\eta} + \sigma_i)^{-1}X. \tag{8.85}$$

Now, substituting (8.84) and (8.85) into (8.82), we have

$$I - \overline{\eta}K_k(t)CB = X^{-1}\mathrm{diag}(\sigma_i)XX^{-1}\mathrm{diag}(\overline{\eta} + \sigma_i)^{-1}X$$
$$= X^{-1}\mathrm{diag}\left(\sigma_i/(\overline{\eta} + \sigma_i)\right)X. \tag{8.86}$$

Finally, since $\sigma_i > 0$, the spectral radius is less than 1. ∎

These results are summarized in the following theorem:

Theorem 8.28. *Under the intermittent measurement environment, η, with mean $\overline{\eta}$, the ILC learning gain determined by (8.75) guarantees $P_{11,k} \to 0$ as $k \to \infty$.* ∎

Proof. By Lemma 8.27, the proof is direct. ∎

Remark 8.29. From the definition of $\overline{\eta}$, we know that when $\overline{\eta} = 1$, there is no intermittent measurement, while when $\overline{\eta} = 0$, all measurements are lost. From (8.86), when $\overline{\eta} = 1$, the spectral radius of $I - \overline{\eta}K_k(t)CB$ is the smallest. So, we can conclude that without intermittent measurement, the best convergence is achieved. However, from (8.86), as far as $\bar{\eta} \neq 0$, still the spectral radius of $I - \bar{\eta}K_k(t)CB$ is less than 1. So, in ILC, the convergence is always guaranteed even if most of the measurements are lost (in other words, as far as there is a part of the data measurement available, whatever amount it is).

Remark 8.30. In (8.75), $K_k(t)$ depends on P, which is the expectation of XX^T. Actually, $K_k(t)$ is calculated based on past information at every iteration. Similarly to [395], we can develop an algorithm for updating $K_k(t)$. To avoid confusion, we provide an algorithm in this remark. Defining $E[\delta x_k(t+1)\delta x_k(t+1)^T] = P_{22,t+1}$ and $E[\delta x_k(t)\delta x_k(t)^T] = P_{22,t}$, from (8.71), we have

$$P_{22,t+1} = BP_{11}B^T + AP_{22}A^T + Q_{11}. \tag{8.87}$$

Let us assume that P_{11}, when $k = 0$, is available and P_{22}, when $t = 0$, is also available. Then, the following algorithm can be developed:

- From (8.87), when $k = 0$, we calculate $P_{22,t+1}$.
- Use (8.75) for updating $K_k(t)$.
- Calculate $u_{k+1}(t)$ using (8.67).
- Use (8.78) to update P_{11}.
- Repeat the whole process (i.e., $k - k + 1$).

Note however, that we are not quite done, as it is necessary to analyze the convergence of P_{22} as well. The following theorem provides the necessary result:

Theorem 8.31. *If there is no initial resetting error at every iteration, as the number of iterations goes to infinity, i.e., $k \to \infty$, then $P_{22} = E[\delta x_k(t)\delta x_k(t)^T]$ goes to*

$$\sum_{i=0}^{t-1} A^{t-1-i}Q_{11} \sum_{i=0}^{t-1} (A^T)^{t-1-i}. \tag{8.88}$$

■

Proof. From (8.68), and after some manipulations, we obtain:

$$\delta x_k(t) = A^t \delta x_k(0) + \sum_{i=0}^{t-1} A^{t-1-i}[B\delta u_k(i) - w_k(i)]. \tag{8.89}$$

Since there is no initial resetting error, $\delta x_k(0) = 0$, so we have

$$E[\delta x_k(t)\delta x_k(t)^T] = \sum_{i=0}^{t-1} A^{t-1-i}E\left[(B\delta u_k(i) - w_k(i))\right.$$
$$\left. \times (\delta u_k(i)^T B^T - w_k(i)^T)\right] \times \sum_{i=0}^{t-1} (A^T)^{t-1-i}$$

$$= \sum_{i=0}^{t-1} A^{t-1-i} \left[BE(\delta u_k(i)\delta u_k(i)^T)B^T \right.$$

$$\left. + E(w_k(t)w_k(t)^T) \right] \sum_{i=0}^{t-1} (A^T)^{t-1-i}. \qquad (8.90)$$

Therefore, from Theorem 8.28, since $E(\delta u_k(i)\delta u_k(i)^T) = 0$ as $k \to \infty$, the proof is completed. ∎

Remark 8.32. In Theorem 8.31, it is shown that P_{22} converges to a fixed value as the number of iterations increases. It is observed that the final converged value of P_{22} depends on A and Q_{11}. If there is no noise, then $P_{22} \to 0$.

When $Q_{11} \neq 0$, we can write (8.88) as:

$$\text{vec}(P_{22}) = \sum_{i=0}^{t-1} (A^i \otimes A^i)\text{vec}(Q_{11}) \qquad (8.91)$$

where $\otimes$ is the Kronecker product, $\text{vec}(P_{22}) = [(P_{22})_1^T, (P_{22})_2^T, \ldots, (P_{22})_n^T]^T$, and $\text{vec}(Q_{11}) = [(Q_{11})_1^T, (Q_{11})_2^T, \ldots, (Q_{11})_n^T]^T$, where $(P_{22})_j$ is the jth column vector of matrix P_{22} and $(Q_{11})_j$ is the jth column vector of matrix Q_{11}. Furthermore, using the property $A^2 \otimes A^2 = (A \otimes A)(A \otimes A)$, we can see that the boundedness of P_{22}, on the time domain, depends on eigenvalues of $A \otimes A$ (not the eigenvalues of A).

8.3.3 Concluding Remarks

In this section, we provided a synthesis method for ILC systems subject to intermittent measurements. Although the algorithm requires using full information about the A matrix, the result is surprising because as long as some measurements are available, the convergence of the ILC system is guaranteed. Our result also can be interpreted in other way, compared with existing intermittent estimation theories [463, 281, 273, 272, 271, 552, 411, 482, 412]. In these works, there are critical data dropout rates, above which the system is not stable any more. However, from our intermittent Kalman filtering scheme enhanced by ILC, we find that there is no critical data dropout rate as far as the control gain is updated by (8.75) on the iteration domain. Intuitively, this surprising result can be understood in the following way: Even though there are lots of data dropouts in the output measurements, if we update the control signals on the repetitive iteration domain, the effect of data loss can be compensated for eventually. Therefore, the authors believe that there is a huge advantage to using ILC scheme when there are data dropouts in the output measurement, assuming the system is operated repetitively. In future work, we consider the design of the learning gain matrix without using full information of the A matrix and will also consider the problem of data dropout in the forward control signal in an NCS system.

8.4 Chapter Summary

In this chapter, we proposed an analysis method to find the baseline error for stochastic ILC processes in the iteration domain. In Section 8.1, the baseline error was analytically calculated using a fixed learning gain matrix, while in Section 8.2, the baseline error was calculated using an iteration-varying learning gain matrix, for the case of iteration-varying model uncertainty. In Section 8.3 we considered ILC algorithms in a networked control system that are subject to intermittent loss of measurement data. The main contributions of this chapter can be briefly summarized as follows:

- In Section 8.1, the analytical solution of an algebraic Riccati equation was used to establish the existence of and to compute the unique positive-definite steady-state error covariance matrix associated with a stochastic ILC system. Based on this result, we also established a condition for monotonic convergence of the stochastic ILC process. Existing works in the literature have focused on iteration-varying gain design in the time domain. Our contributions go beyond these results by designing a fixed learning gain matrix in an off-line manner. Our results are practically important since it is now possible to avoid undesirable overshoot during the transient period of the ILC convergence.

- In Section 8.2, using interval concepts, we found a monotonic convergence condition for iteration-varying uncertain ILC systems. Based on this analysis, in Theorem 8.17 we provided a method for designing iteration-varying learning gain matrices. The contribution of this section over an existing work [64], which only established the baseline error of a deterministic system, is that our result allows us to analytically compute the baseline error of the uncertain ILC system.

- In Section 8.3, the learning gain designed by (8.75) guaranteed the convergence of the mean error of control signal, which implies that the desired trajectory is achieved in a stochastic sense, even with intermittent measurements. This result can be variously used for robust ILC system design or robust intermittent Kalman filtering design. As a main theoretical contribution, we showed that there is no critical data dropout rate when the control gain is updated by (8.75) on the iteration domain.

9

Conclusions

In this monograph we have presented a systematic approach to the analysis and design of iterative learning controllers (ILC) for several classes of uncertain plants, with an emphasis on robustness and monotonic convergence in the iteration domain. Throughout, we used the super-vector iterative learning control (SVILC) framework. The monograph considered robust monotonic stability analysis and synthesis for the case of interval plants, H_∞-based ILC for iteration-domain frequency uncertainty, stochastic ILC when the disturbances and noise are iteration-domain dependent, and the problem of ILC when there is intermittent data dropout. We also provided the solution to three fundamental interval computation problems.

We began in Chapters 1 and 2 with an introduction to the ILC paradigm and its literature, respectively, and continued in Chapter 3 with a discussion of monotonic convergence for higher-order ILC (HOILC) in the SVILC framework. Following the introduction, the next part of the monograph was focused on systems with parametric interval uncertainties. In Chapter 4, vertex Markov matrices were used for finding an exact monotonic-convergence condition for first-order ILC systems (FOILC). We also solved a maximum singular value problem for interval matrices to find an l_2 monotonic-convergence condition for interval ILC processes. Of particular interest, it was shown that the robust stability condition for an interval polynomial matrix can be directly used to check the robust stability of HOILC systems. In Chapter 5 and Chapter 6, robust synthesis problems were discussed. The maximum stability radius was found using a discrete Lyapunov inequality equation and the interval model conversion concept was introduced to estimate the boundaries of the Markov parameters from an interval plant. It was also shown that the main issue in the interval model conversion problem is to find the power of interval matrix. Thus, in Appendix D, we developed an algorithm for computing such powers. The third part of the monograph considered iteration-domain uncertainty other than parametric interval uncertainty. Chapter 7 was dedicated to H_∞ SVILC, where it was shown that iteration-varying external disturbances, iteration-varying model uncertainties, and iteration-dependent stochastic noise could be accounted for together on the iteration axis, although the discussion was limited to an asymptotic stability. In Chapter 8, the baseline error of ILC systems was analytically established using a Kalman filter and an iteration-varying learning gain matrix. For an analytical Kalman filter convergence condition, an algebraic Riccati equation was used, and for the iteration-varying learning gain matrix design, a Wiener filter was used. Finally, we presented results for the problem of ILC design when there is intermittent data dropout along the iteration axis. We point out that in addition to collecting our solutions to three fundamental interval computation problems, the appendices also include a taxonomy of the ILC literature from 1998 to 2004.

The results presented in Chapter 4 to Chapter 8 can be summarized by a table. Based on the 1-dimensional representation of the 2-dimensional ILC system as depicted in Figure 9.1, Table 9.1 has been created to show the results presented in the monograph. In Figure 9.1, the symbol H is the ILC system plant, which we have called the Markov matrix, C is the ILC controller, which can include current-cycle feedback and incorporates the iteration-domain update, D and D_o are disturbances on the iteration domain, N is measurement noise on the iteration domain, and Y_r and Y are the reference trajectory and output trajectory, respectively. Note that in this figure, the feedback loop is in the iteration domain, which is the basic feature of the super-vector approach. In Table 9.1, I.V. stands for iteration varying, z represents the delay operator in the discrete-time domain, and w represents the delay operator in the discrete-iteration domain. In this table, superscript ‡ is used to note the results presented in this monograph, and superscript † is used to indicate topics for future work that have not been completely studied in the existing literature.

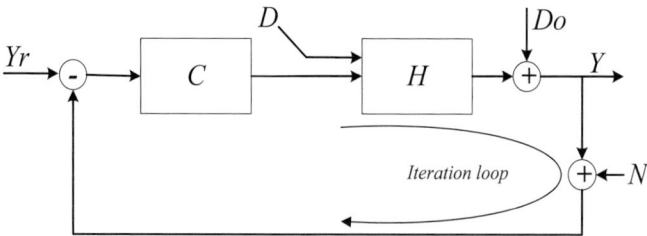

Fig. 9.1. Iterative learning control block diagram in the multi-input, multi-output super-vector framework.

To summarize, we repeat the list of original contributions of the monograph:

- Conditions for robust stability in the iteration domain were provided for parametric interval systems.
- Techniques for converting from time-domain interval models to Markov interval models were given.
- Methods were given to design monotonically-convergent ILC systems under parametric interval uncertainties and/or stochastic noise.
- Robust H_∞ ILC was designed on the iteration domain, taking into account three different types of uncertainties: iteration-variant/invariant model uncertainty, external disturbances, and stochastic noise.
- The baseline error of the ILC process was analytically established, which provides a way to design the ILC learning gain matrix off-line.
- Solutions for three fundamental interval computational problems were given: robust stability of an interval polynomial matrix system, the power of an interval matrix, and the maximum singular value of an interval matrix.

Table 9.1 suggests future research topics in the context of the SVILC framework we have used in this monograph. We conclude by offering additional suggestions for future research efforts that go beyond the SVILC aproach.

- *Linear ILC versus nonlinear ILC.* We believe both linear and nonlinear ILC problems are interesting and important. Yet in this monograph we have focused only on linear ILC, although there exists a significant literature related to nonlinear ILC. It is possible to justify our focus by noting that ILC is a finite-duration, multi-pass system with resetting of initial conditions. As such, we can use linear approximation along the finite-time axis while treating the nonlinear effects as iteration-invariant disturbances, which are readily rejected using the traditional linear ILC framework, which works even for linear time-varying cases where H is not Toeplitz. Thus, the linear framework addresses many of the key problems found in practice, since most existing systems, although inherently nonlinear, do not have a finite escape time and can be stabilized (in iteration) by the ILC process. However, we admit that further advances can be obtained by taking a more deliberate nonlinear point of view and we expect that nonlinear ILC research may be more important in the future, both with respect to ILC-based control of nonlinear plants and with respect to nonlinear ILC update laws.
- *Theory and application.* Linear ILC plays an important role in high tracks-per-inch hard disk drive servo systems. Other notable and successful applications include robotic welding and run-to-run batch processing in the chemical and semiconductor manufacturing industries. We believe ILC is in general a matured methodology that can be routinely applied in control engineering practice. More emphasis should be put on pursuing new ILC

Table 9.1. Summary Table of SVILC

Y_r	D	N	C	H	Framework
Constant	0	0	$\Gamma(w-1)^{-1}$	H_p	Classic "Arimoto" approach (AS)
Constant	0	0	$\Gamma(w-1)^{-1}$	$H_p(w)$	"Owens & Edwards" multipass/linear repetition (AS)
$Y_r(z)$	$D(w)$	0	$C(w)$	H_p	General stability and performance problems (AS)
$Y_r(z)$	$D(z)$	0	$C(w)$	$H(z)$	Time-axis-based frequency-domain analysis of ILC (AS)
$Y_r(z)$	$D(w)$	0	$C(w)$	$H(w)+\Delta(w)$	Frequency uncertainty on the iteration domain (AS)[†]
$Y_r(z)$	0	$w(t),v(t)$	$C(w)$	H_p	Stochastic ILC (AS, MC) on iteration domain[‡]
$Y_r(z)$	0	$w(t),v(t)$	$\Gamma(w-1)^{-1}$	H_p	Least quadratic ILC (AS)
$Y_r(z)$	0	0	$\Gamma(w-1)^{-1}$	H^I	Interval ILC (MC)[‡]
$Y_r(z)$	$D(z)$	$w(t),v(t)$	$C(w)$	$H(z)+\Delta H(z)$	Time domain H_∞ (AS)
$Y_r(z)$	$D(w)$	$w(k,t),v(k,t)$	$C(w)$	$H(w)+\Delta H(w)$	Iterative domain H_∞ (AS)[‡]
$Y_r(z)$	0	$w(k,t),v(k,t)$	$\Gamma(k)$	$H_p+\Delta H(k)$	Iteration-varying uncertainty (MC)[‡]
$Y_r(z)$	0	$\tilde{H}$	Γ	H_p	Intermittent measurement (AS or MC)[†]
$Y_r(w)$	0	0	$\Gamma(w-1)^{-1}$	H_p	Periodic iteration-varying reference (asymptotical)
$Y_r(w)$	0	0	$\Gamma(w-1)^{-1}$	H_p	Periodic iteration-varying reference (monotonic)

applications and the new theoretical issues that will emerge from these new applications.

- *Spatially and temporally evolving dynamic systems.* Using ILC for systems governed by partial differential equations (PDEs) is not well understood. Further, in practice, for most sensors modeled by PDEs, it is usually not possible to also have continuously distributed measurements and control actions. Thus, we must consider the problem when the sensors and actuators are configured for point-wise sensing/actuation, filament-type sensing/actuation, zonal-type sensing/actuation, or boundary-type sensing/actuation. Moreover, we can consider cases where the sensors and actuators are collocated or non-collocated, movable or static, communicating with each other or not, etc. These observations suggest numerous new directions for ILC research.

- *Integer-order versus non-integer-order dynamic systems.* We believe that ILC for fractional-order dynamic systems (polymers, piezo-materials, silicon gel, etc.) is an interesting new area, which involves non-integer calculus and differential equations having non-integer-order derivatives.

- *Fractional order signal-processing-enhanced ILC.* Related to non-integer-order dynamic systems, in recent years, fractional-order signal processing (FOSP) has become a very active research area, due to the demand on precise analysis of long-range dependence and self-similarity in time series, such as financial data, communications networks data, traffic data, water demand, lake volume fluctuation, heart rate variation, bio-corrosion signals, electrochemical noise, etc. We believe joint time–frequency-domain techniques such as wavelets and time–frequency analysis (TFA) can be used in ILC to further reduce the baseline error. Further, since the fractional-order Fourier transform (FrFT) has a strong link with TFA, we believe that in certain applications, FOSP will offer additional benefits in ILC.

- *ILC over networks.* ILC in the network control system (NCS) setting (telepresence, tele-training, etc.) could provide benefits for some classes of systems, as discussed in Chapter 8. But, in the case of NCS, we face the problem of intermittent sensing and actuation and thus intermittent learning updating. The problem of asynchronous ILC can also arise in this setting. Such network-induced issues in ILC suggest a variety of research topics.

- *Cooperative ILC.* Recently there has been significant research activity on the problem of over-populated or densely distributed fields of sensors and actuators, possibly networked and possibly mobile, each with dynamic neighbors under uncertain communication topologies. In this context one can pose the problem of iterative learning-based consensus building for cooperative iterative learning control. As memory and communication get cheaper and cheaper we can envision the concept of "ubiquitous collaborative iterative learning control."

Appendices

A

Taxonomy of Iterative Learning Control Literature

A.1 Taxonomy

In Appendix A, we categorize the ILC literature into two different parts in order to highlight the overall trends in ILC research. The first part is related to the publications that focus on ILC applications and the second part is related to the publications that consider theoretical developments. However, it is in fact very difficult to separate the literature into these two broad groupings, thus the categorizations given in this appendix are based on authors' subjective decisions. Also, note that for this categorization effort, the first author read the abstracts of all the papers indicated. If it was possible to understand the main approach and idea of the paper from the abstract, then the paper was categorized on that basis. However, when the author could not understand how to categorize a paper from its abstract, then other parts of the paper were read in order to decide on the correct category for the paper. As already commented, however, this categorization is still based on subjective decisions and is not a technical development. Also, as mentioned in Chapter 2, the literature search for our taxonomy covers only the publications between 1998 and 2004, because the literature before 1998 was surveyed and classified in [299].

A.2 Literature Related to ILC Applications

In [299], ILC literature dealing with applications was categorized as "robotics" and "applications." In "robotics," detailed categories were given as "elastic joints," "flexible links," "cartesian coordinates," "neural networks," "cooperating manipulators," "hybrid control," and "nonholonomic." In applications, detailed categories were given as "vehicles," "chemical processing," "mechanical/manufacturing systems," "nuclear reactor," "robotics demonstrations," and "miscellaneous." In this section, we began by trying to follow the above categories, but found it difficult to restrict all the publications between 1998 and 2004 into the categories given in Table 4.2 of [299]. Thus, we made more detailed categories, including the topics "Robots," "Rotary systems," "Batch/factory/chemical processes," "Bio/artificial muscle," "Actuators," "Semiconductors," "Power electronics" and "Miscellaneous," and in each category we provide further sub-categories.

A.2.1 Robots

In [299], robotics was the most active area of ILC application. Since 1998, this continues to be the case. Robotic applications of ILC have included:

- General robotic applications, including rigid manipulators and flexible manipulators [109, 166, 449, 408, 175, 329, 222, 422, 448, 21, 223, 36, 213, 533, 203, 542, 541, 174, 555, 109, 225, 319, 204, 240, 95].
- Mechatronics design [462, 487].
- Robot applications with adaptive learning [423].
- Robot applications with Kalman filters [323]
- Impedance matching in robotics [316, 472, 49, 32].
- Table tennis [289].
- Underwater robots [403, 404, 234].
- Acrobat robots [535, 483].
- Cutting robots [224].
- Mobile robots [304, 71].
- Gantry robots [183].

A.2.2 Rotary Systems

Rotational motion is generally disturbed by position-dependent or time-periodic external disturbances. Thus, control of rotary systems is a good candidate for ILC applications. Papers related to this area include:

- The vibration suppression of rotating machinery [266].
- Switched reluctance motors (SRM) [402, 401, 399, 400].
- Permanent-magnet synchronous motors (PMSM) [250, 500, 373, 262, 375, 249, 499, 374].
- Linear motors [441].
- (Ultrasonic) induction motors [398, 285, 284].
- AC servo motors [407].
- Electrostrictive servo motors [206].

A.2.3 Batch/Factory/Chemical process

The number of ILC applications in process control has increased significantly since 1998. The literature includes:

- Tracking control of product quality in agile batch manufacturing processes [493, 256].
- Chemical reactors [294, 293, 264].
- Water heating systems [497].
- Laser cutting [459].
- Chemical processes [151, 495, 45, 189, 445].
- Batch processes [494, 496, 88].
- Industrial extruder plants [352].
- Problem of a moving liquid container [170, 383].

A.2.4 Bio/Artificial Muscle

Bioengineering and biomedical applications are not yet a popular ILC application area, but slowly the number of applications is increasing, as evidenced by the following:

- Biomedical applications such as dental implants [211, 212].
- FNS applications [117, 489].
- Human operators [14].
- Artificial muscle [205].
- Pneumatic systems [53].
- Biomaterial applications [469].

A.2.5 Actuators

ILC applications to non-robotic/non-motor actuators are closely related to the mechanical hard-nonlinearity compensation problem. Related publications are as follows:

- A proportional-valve-controlled hydraulic cylinder system [50].
- Electromechanical valves [198, 361, 199].
- The hysteresis problem of a piezoelectric actuator [280].
- Linear actuators [261, 444].

A.2.6 Semiconductor

It is quite interesting to see that ILC is widely applied in the semiconductor production process. Between 2001 and 2003, the following literature was published in semiconductor applications of ILC: [537, 255, 94, 113, 89, 112, 87, 392, 111, 110]. For a more detailed discussion of the application of ILC to semiconductor manufacturing processes, refer to [114].

A.2.7 Power Electronics

Examples of ILC applications to electrical power systems can be found in the following:

- Electronic/industrial power systems [550, 549, 551, 478].
- Inverters [2, 41].

A.2.8 Miscellaneous

Many miscellaneous applications of ILC are described in the following papers:

- Traffic [201].
- Magnetic bearings [101, 76].
- Aerospace [67, 73].
- Linear accelerators [248].
- Dynamic load simulators [477].
- Hard disk drives [512, 513].
- Temperature uniformity control [254, 292, 258].
- Visual tracking [274].
- Quantum mechanical systems [364].
- Piezoelectric tube scanners [197].
- Smart microwave tubes [1, 405].

Figure A.1 plots the number of papers focused on the use of ILC in applications. As shown in this figure, ILC has most-dominantly been applied to the area of robotics. However, notably, ILC has also been widely used in rotational motion control systems, in the process control industry, and for semiconductor manufacturing processes.

We also note that to check the practical uses of ILC, we searched United States patent abstracts using the keywords "Iterative" AND "Learning." [1] From this search, we found ILC-related patents in motor control [371], process control [168], disk drive control [72], and network communication [230].

A.3 Literature Related to ILC Theories

Since the spectrum of the theoretical developments is so broad and individual papers often treat several different topics, assigning a given paper to a specific category can be quite subjective. The approach in this monograph is to try to separate papers that considered ILC as a specific topic from those that connected ILC analysis with other control theory topics. The general categories are defined as "General (Structure)," "General (Update Rules),"
"Typical ILC Problems," "Robustness," "Optimal and Optimization," "Adaptive," "Fuzzy and Neural," "Mechanical Nonlinearity Compensation," "ILC

---

[1] http://patft.uspto.gov/netahtml/search-bool.html

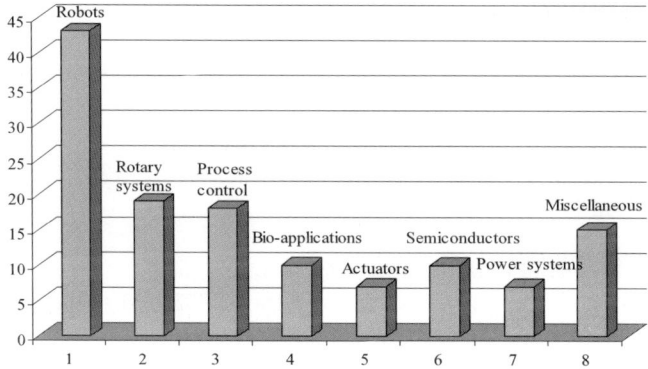

Fig. A.1. Publication numbers for the application-focused ILC literature

for Other Repetitive Systems and Control Schemes," and "Miscellaneous." The first three categories are related to unique ILC problems (i.e., ILC's own issues not related to other control theories). The next four categories (robust, optimal, adaptive, fuzzy/neural) are for papers that combine or use results from these specific fields to advance the theoretical developments of ILC. The next two categories consider special cases where ILC has been applied to develop theoretical solutions to these special problem classes (mechanical non-linearity and repetitive control) and the final category collects miscellaneous contributions.

A.3.1 General (Structure)

In this category, we include literature related to "ILC structure," "convergence analysis," "stability analysis," and "basic theoretical works."

- Structure [346, 351, 363, 180, 454, 347].
- Equivalence of ILC to one-step minimum prediction control or feedback control [302, 466, 162, 345, 163, 164, 161, 165].
- Analysis in the point of passivity (dissipativity) [31, 26, 25, 33, 315, 24].
- Analysis in the point of positivity [182, 145].
- Divergence observation [277].
- Steady-state oscillation condition and its utilization [238].
- Strongly positive system [11, 12].

A.3.2 General (Update Rules)

In this category, we include literature that discusses "ILC update rules" and their "performance comparisons."

- Update rules such as D-type ILC, P-type ILC, I-type ILC, PD-type ILC, and PID-type [502, 409, 63, 536, 410, 547].

- Fractional ILC [60].
- Using current-cycle feedback [369].
- Anticipatory ILC [473].
- Update in Hilbert space [35, 34].
- Performance guaranteed ILC, convergence speed improvement, or performance improvement [504, 195, 515, 509, 510, 70, 492, 128, 246, 130].
- Linearization [190].
- Automated tuning [278].
- Comparison of ILC update rules [498, 503, 320, 514].
- Discussion on convergence and/or robustness [331, 251, 335, 517].

A.3.3 Typical ILC Problems

In this category, we include ILC problems such as nonminimum phase, initial condition reset, higher-order approach, 2-D analysis, and frequency-domain analysis. From Table 4.1 of [299], it is shown that these typical ILC problems had been popularly studied before 1997. But, we can observe that many publications are still devoted to these topics.

- Nonminimum phase and/or noncausal filtering [158, 216, 157, 333, 218, 318, 413, 286, 414, 154, 242, 92, 219, 241, 465].
- Inverse model-based or pseudo-inverse-based ILC [156, 155, 317].
- Initial reset condition [426, 353, 75, 355, 131, 431, 429, 430, 435, 438, 434].
- Higher-order ILC [6, 74, 239, 516, 365, 328].
- 2-D approach/analysis [129, 123, 301, 99, 124, 385, 147, 348, 148, 135, 169, 145, 122].
- Frequency-domain analysis and/or synthesis based on frequency-based filtering [377, 127, 327, 265, 326, 330].

A.3.4 Robustness Against Uncertainty, Time Varying, and/or Stochastic Noise

This category includes robustness problems such as disturbance rejection, stochastic affects, H_∞ approaches, etc. Arif et al. [17] used the following update rule: $u_{k+1}(t) = u_k(t) + \Gamma_1 \dot{e}_k(t) + \Gamma_2 \dot{e}_{k+1}(t)$, where $\dot{e}_{k+1}(t)$ is the predicted error, to improve the ILC convergence speed for time-varying linear systems with unknown but bounded disturbances. In [455], time-periodic disturbances and unstructured disturbances were compensated using a simple recursive technique that does not use Lyapunov equation (refer to [508] for disturbance compensation using Lyapunov functions). For general ideas about robust ILC, refer to [296] for the linear case and see [475, 529, 77, 368, 520, 450, 452] for the nonlinear case. Other related papers include:

- Disturbance rejection with feedback control [86].
- Disturbance rejection with an iteration-varying filter [325].

- Nonlinear stochastic systems with unknown dynamics and unknown noise statistics [55].
- Stochastic ILC [397, 396, 395, 54], and with error prediction [17].
- With measurement noise [313, 324].
- H_∞ approach [360].
- μ-synthesis approach to ILC [115, 116].
- Model-based ILC [23, 382].
- ILC based on the backstepping idea [455].
- Polytope uncertainty approach to ILC [267].

A.3.5 Optimal, Quadratic, and/or Optimization

Optimal ILC is considered one of the main ILC theoretical areas and it has a well-established research history. Norm-optimal ILC is due to [143] as commented in [185]. Recently there have been several different quadratic cost function-based ILC algorithms. Papers in this category include:

- Optimal ILC [9, 181, 185, 186, 187, 342, 394, 341, 458, 5].
- Optimization-based ILC [188, 171].
- Linear quadratic optimization-based method [340, 167].
- Quadratic cost function-based method [91, 237, 253].
- Numerical optimization [288].

A.3.6 Adaptive and/or Adaptive Approaches

Adaptive control-based ILC is very popular and many theoretical works in ILC are related to Lyapunov functions and/or adaptive control concepts. In this category, we only include the literature which focuses on purely theoretical adaptive ILC.

- General works [424, 133, 134, 349, 539, 295, 79, 80, 96, 56, 132, 350, 322].
- Model reference [260].
- Model reference with basis functions [366, 485].
- State-periodic adaptive learning control [3].

A.3.7 Fuzzy or Neural Network ILC

In the ILC literature, it has been shown that learning gains can be determined from neural network or fuzzy logic schemes [298]. Specific results include:

- Fuzzy ILC or fuzzy ILC for initial setting [78, 479, 85, 406, 546, 7, 370].
- Feedforward controller (LFFC) using a dilated B-spline network [65, 464].
- Artificial neural networks/neural networks, or ILC application to neural networks [534, 446, 83, 193, 208, 221, 97, 100, 227, 226, 275, 490, 523, 442, 480, 84].

A.3.8 ILC for Mechanical Nonlinearity Compensation

Many ILC publications show that mechanical hard nonlinearities can be compensated for successfully if they have some sort of periodicity in the time, state, or frequency-domain. The main idea of hard-nonlinearity compensation is to analyze stability in the iteration domain as done in [424].

- ILC without a priori knowledge of the control direction, for non-Lipschitz plants [508, 527, 229].
- ILC with input saturation [505, 519].
- Input singularity [507, 526].
- Deadzone [530, 228, 393].
- Coulomb friction [118, 119, 152, 121, 3].
- Using the Smith predictor for time delay and disturbance systems [207, 511].
- ILC for systems with delay [427, 354, 359].

A.3.9 ILC for Other Repetitive Systems and Other Control Schemes

Though classical control theories have been used for ILC performance improvement, it is also possible to use ILC theory for the performance improvement of other control schemes. Using the general idea of ILC, the performances of several other types of control strategies have been improved, including: repetitive control, PID, optimal control, neural network, etc. [149, 209, 48, 20, 344, 468, 93, 217]; and model-based predictive control [252, 257].

A.3.10 Miscellaneous

Papers that we cannot separate into the categories given above include:

- Different tracking control tasks [506].
- Slowly varying trajectory and/or direct learning control (DLC) for non-repeatable reference trajectory, or DLC for MIMO systems [15, 19, 521, 525, 524, 4].
- LMI-based ILC [144, 421, 150, 146, 386].
- Monotone ILC [62, 303, 306, 310, 308, 357, 358].
- Hamiltonian control systems [140, 138, 139, 141].
- MIMO linear time-varying systems [443].
- Observer-based ILC [450].
- Blended multiple model ILC [451].
- Composite energy function ILC [501, 524].
- Cascaded nonlinear systems [380, 378].
- Nonlinear systems with constraints [57].
- Maximum phase nonlinear systems [90].

- Unknown relative degree [425]; and arbitrary or higher relative degree [81, 436, 545, 22].
- Decentralized iterative learning control [488].
- Internal model-based ILC [453, 61, 64, 457, 528].
- Distributed parameter systems [379, 381].
- ILC with prescribed input–output subspace [173, 176]; with desired input in an appropriate finite-dimensional input subspace [419, 420]; and with bounded input [120].
- Sampled-data ILC [428, 432, 440, 439, 433, 434, 437].
- Experience/information database [16, 18, 15].
- Fourier series-based learning controller [376].
- Learning variable structure control [532].
- With weighted local symmetrical integral feedback controller [69].

Figure A.2 plots the number of papers related to theoretical developments in ILC. As seen in this figure, ILC theory has been advanced by being connected to existing control theories such as robust, adaptive, optimal and neural/fuzzy control. However, the ILC structure and update problems, which are investigated within ILC's own framework, dealing with ILC problems such as the non-minimum phase systems, the initial reset problem, the higher-order issue, 2-D analysis, and convergence/performance improvement, have been more widely studied. Figure A.2 reveals that much research has been devoted to ILC's own theoretical and structural problems. It is also interesting to point out that while other control schemes have been used to help improve ILC, in the same way the ILC concept has been used for the performance improvement of other control schemes.

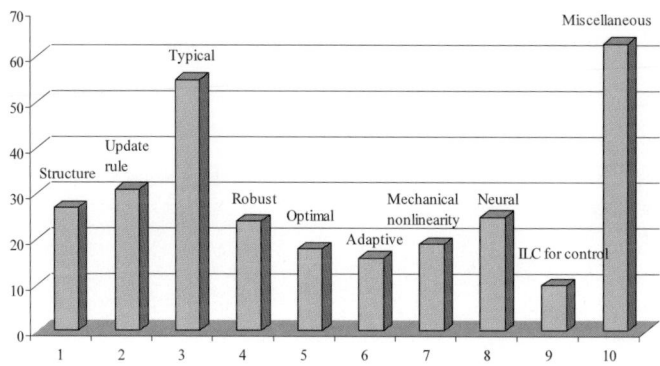

Fig. A.2. Publication numbers for the theory-focused ILC literature

A.4 Discussion

In this appendix we have categorized and discussed the iterative learning control literature published between 1998 and 2004. From the categorization of application-related literature, we have found that ILC applications have been extended from robotics and process control to more specific semiconductor manufacturing and bioengineering applications. However, applications remain dominated by manipulator-based robotics, rotary systems, and process control problems, which are basically time- or state-periodic in either the desired trajectory or the external disturbances. Although some of the publications have shown that ILC can be used in the areas of aerospace, non-robotic actuator control, biomedical applications, visual tracking, artificial muscles, and other emerging engineering problems, successful industrial applications have not yet been reported in these areas.

From the survey of theory-focused literature, it is seen that ILC theory has been developed in two different areas: research on ILC's own features and research on ILC systems fused with other control theories. Most of the recent theoretical work has been related to performance improvement with various types of uncertainties and/or instabilities. However, although many recent theoretical achievements have provided beautiful mathematical formulations of ILC, much of the theoretical development remains far away from actual application considerations.

B

Maximum Singular Value of an Interval Matrix

In this appendix the maximum singular value of an interval matrix is calculated and the effectiveness of the proposed method is illustrated through numerical examples. Calculations are given for both square and non-square matrices. The maximum singular value of an interval matrix can be used to check the monotonic convergence of uncertain ILC systems.

B.1 Maximum Singular Value of a Square Interval Matrix

To develop our algorithm we make use of Hertz's idea for finding extreme eigenvalues of a symmetric interval matrix [192]. Let us consider a real square non-symmetric interval matrix given as

$$A^I = [a_{ij}^I], \quad a_{ij}^I = [\underline{a_{ij}}, \overline{a_{ij}}], i, j = 1, \ldots, n, \tag{B.1}$$

where a_{ij}^I is an element of the interval matrix A^I, $\underline{a_{ij}}$ is the lower boundary of the interval a_{ij}^I, and $\overline{a_{ij}}$ is the upper boundary of the interval a_{ij}^I. If the lower and the upper boundary matrices are defined as $\underline{A} = [\underline{a_{ij}}]$ and $\overline{A} = [\overline{a_{ij}}]$, respectively, the interval matrix can then be written as $A \in A^I = [A^o - \Delta, A^o + \Delta]$, where the center matrix and the radius matrix are defined, respectively, as

$$A^o = \frac{1}{2}(\overline{A} + \underline{A}); \quad \Delta = \frac{1}{2}(\overline{A} - \underline{A}). \tag{B.2}$$

In fact, the upper boundary of the singular values of an interval matrix can be found as (in descending order) $\sigma_i(A^I) = \sqrt{\lambda_i((A^I)^T \otimes A^I)}$ where $\otimes$ represents multiplication of interval matrices (see Section 4.1), σ is the singular value, and λ is the eigenvalue. However, as commented in [106], the results of this method will be quite conservative. Thus, we propose an exact calculation that will not be conservative.

To begin, consider the following relationship between singular values and eigenvalues:

$$\sigma_i(A) = \text{Positive}\left(\lambda_i \begin{bmatrix} 0 & A^T \\ A & 0 \end{bmatrix}\right), \tag{B.3}$$

where Positive$(\cdot)$ considers only the positive part of $(\cdot)$. Define

$$H = \begin{bmatrix} 0 & A^T \\ A & 0 \end{bmatrix}.$$

Obviously, H is a symmetric matrix. Including interval uncertainties, let us use also define the symmetric interval matrix

$$H^I = \begin{bmatrix} 0 & (A^I)^T \\ A^I & 0 \end{bmatrix}.$$

Now, we make use of the existing result from [192] to find the maximum singular value of A^I. In what follows, the main idea and results are briefly summarized. From [192], since H and H^I are symmetric matrices, we have the relationship:

$$\lambda = x^T H x = x^T \begin{bmatrix} 0 & A^T \\ A & 0 \end{bmatrix} x, \tag{B.4}$$

where x is an eigenvector corresponding to λ and $x^T x = 1$. Let us divide x into two parts given as $x^T = [y^T, z^T]$. Then, $y_i = x_i$, $i = 1, \ldots, n$ and $z_i = x_{n+i}$, $i = 1, \ldots, n$, and from (B.4), we obtain $\lambda = 2\left(\sum_{i=1}^{n} \sum_{j=1}^{n} a_{ji} y_i z_j\right)$. Therefore, the value of λ depends on signs of y_i and z_j. That is, the maximum of λ occurs at one of the vertex points of a_{ij}, which is given as

$$a_{ij} = \begin{cases} a_{ij} = \overline{a_{ij}} & \text{if } y_i z_j \geq 0, \\ a_{ij} = \underline{a_{ij}} & \text{if } y_i z_j < 0. \end{cases} \tag{B.5}$$

Now, since y and z are length-n vectors, we have total number of 2^n different sign patterns for y and 2^n different sign patterns for z. For example, when $n = 3$, the sign patterns of y and z could be $+++$, $++-$, $+-+$, $+--$, $-++$, $-+-$, $--+$, $---$. In this case, we have a total of $2^3 \times 2^3 = 64$ combinations as shown in Table B.1 and Table B.2. However, Table B.1 and Table B.2 produce the same vertex matrices set for A^I. Therefore, for our purpose, it will be enough to check a total of 2^5 vertex matrices corresponding to Table B.1. These vertex matrices can be found easily. For example, in Table B.1, for the sign pattern $+--$ of y and for the sign pattern $+-+$ of z, the sign of the vertex matrix is generated from zy^T as

$$\begin{bmatrix} + \\ - \\ + \end{bmatrix} [+ \quad - \quad -] = \begin{bmatrix} + & - & - \\ - & + & + \\ + & - & - \end{bmatrix}, \tag{B.6}$$

Table B.1. 32 Sign Patterns with $\text{Sign}(y_1) = +$ for a 3×3 Matrix

y	z	y	z
	$+++$		$+++$
	$++-$		$++-$
	$+-+$		$+-+$
$+++$	$+--$	$+-+$	$+--$
	$-++$		$-++$
	$-+-$		$-+-$
	$--+$		$--+$
	$---$		$---$
	$+++$		$+++$
	$++-$		$++-$
	$+-+$		$+-+$
$++-$	$+--$	$+--$	$+--$
	$-++$		$-++$
	$-+-$		$-+-$
	$--+$		$--+$
	$---$		$---$

which means that the corresponding vertex matrix is

$$
\begin{bmatrix}
\overline{a_{ij}} & a_{ij} & a_{ij} \\
a_{ij} & \overline{\overline{a_{ij}}} & \overline{\overline{a_{ij}}} \\
\overline{a_{ij}} & a_{ij} & a_{ij}
\end{bmatrix}. \tag{B.7}
$$

In the following algorithm, based on the above discussion, for an $n \times n$ interval matrix, a generalized method is developed.

Algorithm B.1. *Algorithm for estimating the maximum singular value of an interval matrix:*

- **Step 1**: *Produce a set of ± 1 vectors with $y_1 = 1$ of length n given by*

$$
Y = \{y \in R^n : y_1 = 1, \ |y_j| = 1, \ \text{for } j = 2, \ldots, n\}.
$$

- **Step 2**: *Produce a set of ± 1 vectors of length n given by*

$$
Z = \{z \in R^n : \ |z_j| = 1, \ \text{for } j = 1, \ldots, n\}.
$$

- **Step 3**: *Make an $n \times n$ diagonal matrix T_y defined by $(T_y)_{ii} = y_i$ and $(T_y)_{ij} = 0$ for $i \neq j$, $i, j = 1, \ldots, n$ where $y \in Y$.*
- **Step 4**: *Make an $n \times n$ diagonal matrix T_z defined by $(T_z)_{ii} = z_i$ and $(T_z)_{ij} = 0$ for $i \neq j$, $i, j = 1, \ldots, n$ where $z \in Z$.*

Table B.2. 32 Sign Patterns with $\mathrm{Sign}(y_1) = -$ for a 3×3 Matrix

y	z	y	z
	$+++$		$+++$
	$++-$		$++-$
	$+-+$		$+-+$
	$+--$		$+--$
$-++$	$-++$	$--+$	$-++$
	$-+-$		$-+-$
	$--+$		$--+$
	$---$		$---$
	$+++$		$+++$
	$++-$		$++-$
	$+-+$		$+-+$
	$+--$		$+--$
$-+-$	$-++$	$---$	$-++$
	$-+-$		$-+-$
	$\vdash$		$--+$
	$---$		$---$

- **Step 5**: *Produce a matrix set*

$$\mathcal{S}^v := \left\{ A_{yz} : A_{yz} = A^o + T_y \Delta T_z, \ \forall \ y \in Y \ \text{ and } \ \forall \ z \in Z \right\}.$$

- **Step 6**: *Find the maximum singular values of all elements of the finite set $\mathcal{S}^v$ and select the largest one as the maximum singular value of the interval matrix A^I.*

B.2 Maximum Singular Value of Non-square Interval Matrix

The results of the preceding section can easily be extended to the general non-square interval matrix case. Let us consider an $m \times n$ interval matrix A^I. Then, H^I is an $(m+n) \times (m+n)$ interval matrix. Now, introducing a length-n vector y and a length-m vector z, using the same procedure as done in the square matrix case, we have $\sigma(A) = 2\left(\sum_{i=1}^{n} \sum_{j=1}^{m} a_{ji} y_i, z_j\right)$. Then, there are total number of 2^{m+n-1} possible combinations of vertex matrices to be considered. For example, for a 3×2 matrix, we have a total of $2^3 \times 2^1$ combinations as shown in Table B.3. In Table B.3, for example, for the sign pattern $+-$ of y and for the sign pattern $+-+$ of z, the sign pattern of the vertex matrix is generated from zy^T to be

Table B.3. 16 Sign Patterns for a 3×2 Non-square Matrix

y	z	y	z
++	$+++$	+−	$+++$
	$++-$		$++-$
	$+-+$		$+-+$
	$+--$		$+--$
	$-++$		$-++$
	$-+-$		$-+-$
	$--+$		$--+$
	$---$		$---$

$$\begin{bmatrix} + \\ - \\ + \end{bmatrix} \begin{bmatrix} + & - \end{bmatrix} = \begin{bmatrix} + & - \\ - & + \\ + & - \end{bmatrix},$$
(B.8)

which means that the corresponding vertex matrix is

$$\begin{bmatrix} \overline{a_{ij}} & a_{ij} \\ a_{ij} & \overline{a_{ij}} \\ \overline{a_{ij}} & a_{ij} \end{bmatrix}.$$
(B.9)

B.3 Illustrative Examples

B.3.1 Example 1: Non-square Case

Let us test the non-square case first. For the non-square case, the following example is adopted from [106]:

$$A \in A^I = \begin{bmatrix} [2,3] & [1,1] \\ [0,2] & [0,1] \\ [0,1] & [2,3] \end{bmatrix}.$$
(B.10)

Using the results given in Section B.2, the maximum singular value of A^I is found to be 4.54306177572459, which is quite close to the value 4.543062 given in [106]. This result shows that the suggested method in this monograph can find the exact (without conservatism) upper boundary of the maximum singular value of an interval matrix. Note that the suggested scheme in this monograph does not require any assumptions.

B.3.2 Example 2: Square Case

Next, for an example with a square matrix and to represent an exception of Deif's method [106], the following center is used:

$$A^o = \begin{bmatrix} -3.33 & -2.24 & 0.06 \\ 1.03 & -0.34 & 1.09 \\ -2.02 & -1.02 & 2.27 \end{bmatrix},$$
(B.11)

with an associate radius matrix taken as:

$$\Delta = \begin{bmatrix} 1.32 & 0.86 & 4.38 \\ 0.84 & 2.97 & 1.42 \\ 1.61 & 3.06 & 0.55 \end{bmatrix}.$$
(B.12)

Using the suggested method, the maximum singular value of A^I is found to be 9.8549, but from Deif's method, it is found to be 9.7408. For demonstration purposes, random tests are performed. Figure B.1 shows the results of a Monte-Carlo-type random test where the maximum singular values of a large number of matrices taken from A^I were computed. In the figure the dash-dot line is the calculated maximum singular value from the suggested method (9.8549) and the solid line is the maximum singular value from Deif's method. Clearly there exist exceptions in the case of Deif's method, while the suggested method bounds the maximum singular values without any exception.

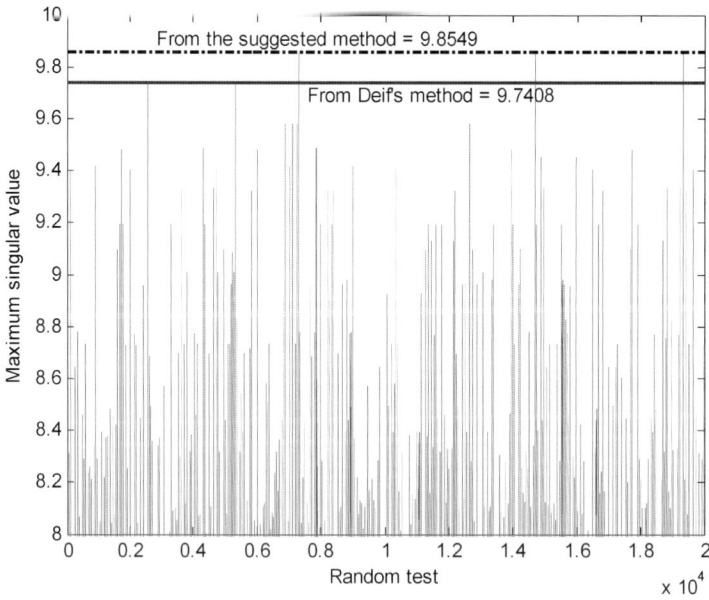

Fig. B.1. Maximum singular values of randomly selected matrices and the calculated maximum singular values from the suggested method (dash-dot line) and from Deif's method (solid line)

B.4 Summary

In this appendix, algorithms for calculating the maximum singular value for square and non-square interval matrices were developed. Using an existing example from [106], whose result was developed based on perturbation theory under some restrictive assumptions, it was verified that the proposed method in this appendix could find the exact maximum singular value. Furthermore, by example it was also shown that the existing method [106] does not find the maximum singular value in some cases while the suggested method finds the maximum singular value without any exception.

C

Robust Stability of Interval Polynomial Matrices

In this appendix, the concept of an interval polynomial matrix system is introduced and a robust stability condition for such systems is derived. This robust stability condition can be used for testing the asymptotic stability of uncertain (interval) higher-order ILC systems.

C.1 Interval Polynomial Matrices

As commented in [372, 270], matrix polynomials (or polynomial matrices) [160, 13] are important in the theory of higher-order vector differential equations, multi-input, multi-output control systems [51], and n-D circuits. For the last two decades, the robust stability problem for polynomial matrices has been steadily studied [372, 243, 232, 476, 481]. In fact, for robust analysis, the interval concept has been quite popular, as shown in [314, 37, 42, 215, 179, 178]. Interval polynomial matrices occur in discrete-time, multivariable problems where physical constants in the plant are subject to interval uncertainty. Under the interval uncertainty concept, after Kharitonov [235] provided an analytical solution for the stability of the continuous interval polynomial, a great amount of literature has been devoted to the study of robustness for interval matrices and interval polynomials. For instance, interval matrices, matrix polytopes, interval polynomial matrices, and polynomial matrix polytopes have been well defined and studied [37, 42, 220, 543, 544, 215, 476, 481]. Also, it has been well known that the stability of continuous interval polynomial matrices or interval matrix polynomials can be checked by Kharitonov polynomials [243, 232, 476, 481]. However, relatively very few research efforts have been devoted to discrete interval polynomial matrices [243] or discrete interval matrix polynomials [233]. Two recent results are Henrion [191], who suggested an LMI condition for the robust stability of polynomial matrix polytopes and polytope type polynomial matrices, and Psarrakos [372], who provided a stability radius for discrete polynomial matrices. In Henrion's work, an LMI condition was used for polynomial matrices when coefficients of the polynomial matrix vary dependently. In Psarrakos's work, it was required to calculate the norm of the inverse of the nominal polynomial matrix, which could be quite conservative. In this appendix, a new analytic Schur stability checking method is developed, which is algebraically simple and less conservative than the existing results. In this newly developed approach, Markov matrices of the polynomial matrix are exploited by using the inverse of the polynomial matrix.

To ensure the consistency of notations and definitions, based on [243, 232, 476, 191, 481], let us repeat the definitions of interval polynomial matrices. When the (i,j)-th element of matrix $P(z)$ is denoted by

$$p_{ij}(z) = a_{ij0} + a_{ij1}z + \cdots + a_{ijm}z^m, \ i,j = 1,\ldots,m, \qquad (\text{C.1})$$

where m is the degree of polynomial $p_{ij}(z)$, matrix $P(z)$ is called a polynomial matrix. When the coefficients of a polynomial lie in intervals like

$$\underline{a_{ijk}} \le a_{ijk} \le \overline{a_{ijk}}, \ k = 0,\ldots,m; i,j = 1,\ldots,n, \qquad (\text{C.2})$$

where n is the dimension of square $P(z)$; $\overline{(\cdot)}$ is the maximum extreme value of $(\cdot)$ and $(\cdot)$ is the minimum extreme value of $(\cdot)$, these polynomial matrices are called interval polynomial matrices (denoted by $P^I(z)$) [243]. Note that polytopic polynomial matrices [37, 476] or polynomial matrix polytopes [191] should be distinguished from interval polynomial matrices. As commented in [191], these polynomial matrix polytopes are linear combinations of a set of given polynomial matrices.

C.2 Definitions and Preliminaries

Let us consider a real monic polynomial matrix of the form:

$$P(z) = I_{m\times m}z^n + A_1 z^{n-1} + \cdots + A_{n-1}z + A_n, \qquad (\text{C.3})$$

where $I_{m\times m}$ is the $m \times m$ identity matrix; the coefficient matrices A_i, $i = 1,\ldots,n$ are $m \times m$ real square matrices, i.e., $A_i \in \mathbf{R}^{m\times m}$; and z is a point in complex plane, i.e., $z \in \mathbf{C}$. The following definitions are then used to discuss the stability of the polynomial matrix $P(z)$.

Definition C.1. [243, 476, 336] The roots λ^* of $\det(P(\lambda)) = 0$ are called eigenvalues of $P(z)$[1]. Thus, when we define a set $S_\lambda = \{\lambda \mid \det(P(\lambda)) = 0\}$, if $\max_{\lambda \in S_\lambda} |\lambda| < 1$, then the polynomial matrix $P(z)$ is robust D-stable. In this appendix, robust D-stability is also called Schur stability without notational confusion.   ∎

---

[1] This notation is not unusual. For a similar discussion, refer to [247, 105, 372, 194, 13].

Definition C.2. *When the elements of a matrix A are intervals such as $a_{ij} \in [\underline{a_{ij}}, \overline{a_{ij}}]$, this matrix is called an interval matrix A^I. The modulus matrix ($|A|_m$) of an interval matrix is defined as*

$$|A|_m = \left[a_{ij}^m : a_{ij}^m \in \max\{|\underline{a_{ij}}|, |\overline{a_{ij}}|, i, j = 1, \ldots, n\} \right],$$

where a_{ij}^m are elements of modulus matrix $|A|_m$. For a non-interval matrix A, we define $|A|_m = |A| = [|a_{ij}|]$, which takes the absolute value of each element of A. ∎

For the derivation of the Schur stability of the interval polynomial matrix $P^I(z)$, the following lemmas are needed.

Lemma C.3. *[51] If $P(z)$ is a real polynomial matrix and $\det(P(z))$ is not identically zero then it is invertible (i.e., non-singular) and its inverse is a real-rational matrix.* ∎

Lemma C.4. *[98, 418] For an $m \times m$ matrix R, if $\rho(R) < 1$ (ρ means spectral radius), then $\det(I \pm R) \neq 0$.* ∎

Lemma C.5. *If $P(z)$ is invertible,[2] then $[P(z)]^{-1}$ can be expanded as $\sum_{k=0}^{\infty} T_k z^{-k}$ (i.e., $[P(z)]^{-1} = \mathrm{span}\{Iz^{-i}, i = 0, \ldots, \infty\}$).* ∎

Proof (of Lemma C.5). By Lemma C.3, there exists $[P(z)]^{-1}$ whose elements are real-rational functions of z, denoted $p_{ij}^{-1}(z) = \sum_{k=0}^{\infty} t_{ij_k} z^{-k}$. That is, each element of $[P(z)]^{-1}$ can be expanded into its Markov parameters. Clearly then, we can write $T_k = [t_{ij_k}]$. ∎

Note that T_k are called the *Markov matrices* of the inverse polynomial matrix $[P(z)]^{-1}$.

Lemma C.6. *[338] For any square matrices, R, T, and V, if $|R|_m \leq V$, then the following inequalities are true:*

$$\rho(RT) \leq \rho(|R|_m |T|_m) \leq \rho(V|T|_m),$$

where the subscript m means the modulus matrix. ∎

C.3 Stability Condition for Interval Polynomial Matrices

The key idea of the suggested method is to utilize Markov matrices in the region $|z| \geq 1$ of the complex plane. The method will be developed based on the matrix determinant. Our results are organized into three subsections. The first two consider the stability of polynomial matrices. The results from these first two subsections are then used in the third subsection to develop our main result.

---

[2] The assumption that $P(z)$ is invertible is practically meaningful. From [51], "all polynomial matrices are invertible for almost all z unless the determinant of $P(z)$ is 0 for all z." Thus, it is the basic assumption that $P(z)$ is invertible.

C.3.1 The Stability of Polynomial Matrices: Part 1

Let us begin this subsection by rewriting the polynomial matrix (C.3) as

$$P(z) = z^{n-1}(zI + A_1 + A_2 z^{-1} + \cdots + A_n z^{-n+1})$$
$$= z^{n-1}(zI + S(z)) = z^{n-1}Q(z), \tag{C.4}$$

where $S(z) := A_1 + A_2 z^{-1} + A_3 z^{-2} + \cdots + A_n z^{-n+1}$, and $Q(z) := zI + S(z)$. By taking the determinant of both sides of (C.4), we have:

$$\det[P(z)] = \det[z^{n-1}I \cdot Q(z)] = \det[z^{n-1}I] \cdot \det[Q(z)]. \tag{C.5}$$

Here, observe that $z = 0$ is the solution such that $\det[z^{n-1}I] = 0$. Furthermore, $z = 0$ is not defined in $Q(z)$, because denominators become zeros. Thus, for the polynomial $Q(z)$, the complex plane without the origin is considered. Define $\mathbf{C}^* = \mathbf{C} - \{0\}$. Then, to determine the stability of $P(z)$ from $\det[Q(z)]$, the following lemmas can be developed:

Lemma C.7. In $\mathbf{C}^*$, from (C.4), $P(z)$ is stable if and only if $Q(z)$ is stable. ∎

Proof. In $\mathbf{C}^*$, $\det[z^{n-1}I] \neq 0$. Hence, only z such that $\det[zI + S(z)] = 0$ makes $\det[P(z)]$ zero. Thus, from the following latent solutions:

$$S_{z^*} = \{z | \det[Q(z)] = 0, \ z \in \mathbf{C}^*\}; \ S_{z^{**}} = \{z | \det[P(z)] = 0, \ z \in \mathbf{C}^*\},$$

the following set equality is true: $S_{z^*} = S_{z^{**}}$. Hence, $P(z)$ is stable if and only if $Q(z)$ is stable. ∎

Lemma C.8. In $\mathbf{C}^*$, the polynomial matrix $Q(z)$ is stable if and only if $\det[Q(z)] \neq 0$, for all $|z| \geq 1$. ∎

Proof. It is certain that, in the complex plane, there exists a z such that $\det[Q(z)] = 0$. Thus, the condition "$\det[Q(z)] \neq 0, \ \forall \ |z| \geq 1$" is equivalent to the condition "there exists a z such that $\det[Q(z)] = 0$ only in the disk of $|z| < 1$." Thus, by Definition C.1, $Q(z)$ is stable. For the "only if" condition, assume that $Q(z)$ is stable. Also assume there exists any z, $|z| \geq 1$ such that $\det[Q(z)] \neq 0$. Then, from $S_\lambda = \{\lambda \mid \det(P(\lambda)) = 0\}$, we have $\max_{\lambda \in S_\lambda} |\lambda| \geq 1$. This contradicts the fact that $Q(z)$ is stable. Hence, $\det[Q(z)] \neq 0, \ \forall \ |z| \geq 1$ is the stability condition. ∎

Therefore, based on the results of Lemma C.7 and Lemma C.8, it is concluded that in $\mathbf{C}^*$, $P(z)$ is stable if and only if $\det[Q(z)] \neq 0$, for all $|z| \geq 1$. The following theorem is then suggested:

Theorem C.9. If $\det(A_n) \neq 0$, then $P(z)$ is stable in $z \in \mathbf{C}$ if and only if $\det[Q(z)] \neq 0$, for all $|z| \geq 1$. ∎

Proof. The proof can be completed by substituting $z = 0$ into $P(z)$. If $z = 0$ is substituted into $P(z)$, then $P(z = 0) = A_n$. Thus, if $\det(A_n) \neq 0$, then $z = 0$ is not a latent solution such that $\det[P(z)] = 0$. Hence, since latent solutions of $\det[P(z)] = 0$ in $\mathbf{C}$ are equivalent to latent solutions of $\det[P(z)] = 0$ in $\mathbf{C}^*$, the stability of $P(z)$ in $\mathbf{C}^*$ means the stability of $P(z)$ in $\mathbf{C}$.   ■

The above theorem shows that the polynomial matrix $P(z)$ is stable if and only if $\det[Q(z)] \neq 0$ for all $|z| \geq 1$ with $\det(A_n) \neq 0$. Next, given this necessary and sufficient condition for the stability of $P(z)$, the remaining problem with respect to the stability of a polynomial matrix is to find an equivalent condition for $\det[Q(z)] \neq 0$. This will be discussed in the following subsection.

C.3.2 The Stability of Polynomial Matrices: Part 2

Now we consider the Markov matrices of polynomial matrices, as a vehicle for discussing the stability of $\det[Q(z)] \neq 0$. Then in the next subsection, based on the boundary condition of the sum of these Markov matrices, a robust stability condition for interval polynomial matrices is developed.

To define the Markov matrices of a polynomial matrix, the following lemma is needed:

Lemma C.10. *If* $\det(P(z))$ *is not identically zero, then the polynomial matrix* $Q(z) = zI + S(z)$ *is nonsingular and* $[Q(z)]^{-1}$ *can be expanded using Markov matrices in* $\mathbf{C}^*$ *as*

$$[Q(z)]^{-1} = \sum_{k=0}^{\infty} T_k z^{n-k-1}.$$

■

Proof. By multiplying $z^{1-n}I$ by $P(z)$, we have:

$$z^{1-n}I \cdot P(z) = (zI + A_1 + A_2 z^{-1} + \cdots + A_n z^{-n+1}) = Q(z). \quad \text{(C.6)}$$

Here, since $P(z)$ is nonsingular from Lemma C.3 and $z^{1-n}I$ is nonsingular in $\mathbf{C}^*$, then clearly $z^{1-n}I \cdot P(z)$ is nonsingular. Also, from Lemma C.5, $[P(z)]^{-1} = \sum_{k=0}^{\infty} T_k z^{-k}$. Thus, the following relationship can be established easily:

$$[zI + S(z)]^{-1} = [z^{1-n}P(z)]^{-1} = z^{n-1}[P(z)]^{-1} = \sum_{k=0}^{\infty} T_k z^{n-k-1}. \quad \text{(C.7)}$$

This completes the proof.   ■

For convenience, let us write $\sum_{k=0}^{\infty} T_k z^{n-k-1}$ as

$$\sum_{k=0}^{\infty} T_k z^{n-k-1} = \sum_{k=0}^{n-2} T_k z^{n-k-1} + \sum_{k=n-1}^{\infty} T_k z^{n-k-1}, \quad \text{(C.8)}$$

and replace the second term of the right-hand side by

$$\sum_{k=n-1}^{\infty} T_k z^{n-k-1} = \sum_{i=0}^{\infty} R_i z^{-i}, \tag{C.9}$$

where $R_0 = T_{n-1}, R_1 = T_n, R_2 = T_{n+1}, \ldots, R_i = T_{n+i-1}$. With this notation, the process for the calculation of the Markov matrices of $Q(z)$ is summarized in the following lemma:

Lemma C.11. *If* $\det(P(z))$ *is not identically zero, the inverse of* $Q(z)$ *is expressed as* $[Q(z)]^{-1} = \sum_{i=0}^{\infty} R_i z^{-i}$ *in which the Markov matrices are calculated by*

$$R_k = -\sum_{i=1}^{k-1} A_i R_{k-i}, \quad k \geq 2, \tag{C.10}$$

with $R_0 = 0$ *and* $R_1 = I$, *and* $A_i = 0_{m \times m}$ *for* $i \geq n+1$. ∎

Proof. From Lemma C.10, the following relationship is given:

$$[Q(z)]^{-1} = \sum_{k=0}^{\infty} T_k z^{n-k-1}.$$

Also, based on Lemma C.10, in $\mathbf{C}^*$, there exists an inverse of $Q(z)$. Thus, the following equalities are true:

$$Q(z)[Q(z)]^{-1} = Q(z) \cdot \sum_{k=0}^{\infty} T_k z^{n-k-1}$$

$$\Leftrightarrow I = Q(z) \cdot \sum_{k=0}^{\infty} T_k z^{n-k-1}$$

$$= \left(zI + A_1 + A_2 z^{-1} + \cdots + A_n z^{-n+1} \right) \left(\sum_{k=0}^{\infty} T_k z^{n-k-1} \right)$$

$$= \left(zI + A_1 + A_2 z^{-1} + \cdots + A_n z^{-n+1} \right)$$

$$\times \left(\sum_{k=0}^{n-2} T_k z^{n-k-1} + \sum_{k=n-1}^{\infty} T_k z^{n-k-1} \right)$$

$$= \left(zI + A_1 + A_2 z^{-1} + \cdots + A_n z^{-n+1} \right)$$

$$\times \left(\sum_{k=0}^{n-2} T_k z^{n-k-1} + \sum_{i=0}^{\infty} R_i z^{-i} \right). \tag{C.11}$$

Here, using the fact that the left-hand side and right-hand side of (C.11) should be equal for all z, after some manipulations it is easy to show that

$T_k = 0_{m \times m}, i = 0, \ldots, n-2$ and the following formula is easily derived for R_k:

$$R_k = -\sum_{i=1}^{k-1} A_i R_{k-i}, \ k \geq 2, \ A_i = 0_{m \times m} \text{ at } i \geq n+1, \qquad \text{(C.12)}$$

with $R_0 = 0$ and $R_1 = I$. Recall that $R_k, k = 0, \ldots, \infty$, are coefficient matrices of the right-hand side of (C.9). Therefore,

$$[zI + S(z)]^{-1} = [Q(z)]^{-1} = \sum_{i=0}^{\infty} R_i z^{-i}, \qquad \text{(C.13)}$$

where R_i are determined in (C.12). ∎

In Lemma C.11, we provided a formula for calculation of the Markov matrices of the inverse of a polynomial matrix. Now, based on Lemma C.11, it is easy to see that $R_k \to 0$ if and only if $Q(z)$ is stable. For more detail, let us change R_k to

$$R_k = -\sum_{i=1}^{k-1} A_i R_{k-i}$$
$$= -A_1 R_{k-1} - A_2 R_{k-2} \cdots - A_n R_{k-n}. \qquad \text{(C.14)}$$

Then, the following relationship is obtained:

$$\begin{bmatrix} R_k \\ R_{k-1} \\ \vdots \\ R_{k-n+2} \\ R_{k-n+1} \end{bmatrix} = \begin{bmatrix} -A_1 & -A_2 & \cdots & -A_{n-1} & -A_n \\ I_{m \times m} & 0_{m \times m} & \cdots & 0_{m \times m} & 0_{m \times m} \\ 0_{m \times m} & I_{m \times m} & \cdots & 0_{m \times m} & 0_{m \times m} \\ \vdots & \vdots & \ddots & \vdots & \vdots \\ 0_{m \times m} & 0_{m \times m} & \cdots & I_{m \times m} & 0_{m \times m} \end{bmatrix} \begin{bmatrix} R_{k-1} \\ R_{k-2} \\ \vdots \\ R_{k-n+1} \\ R_{k-n} \end{bmatrix}. \qquad \text{(C.15)}$$

Denoting the above equation as $\overline{R}_k = \mathcal{C}\overline{R}_{k-1}$, if $\rho(\mathcal{C}) < 1$, then $\|\overline{R}_k\| \to 0$ as $k \to \infty$ which implies $\overline{R}_k \to 0$ as $k \to \infty$. Hence $R_k \to 0$ as $k \to \infty$ if and only if $\rho(\mathcal{C}) < 1$, which is an equivalent condition for the stability of $Q(z)$. Thus, if $\rho(\mathcal{C}) < 1$, we can then calculate the absolute summation of R_k (denoted Σ_{R_k}) according to

$$\Sigma_{R_k} = I + \sum_{k=2}^{\infty} |R_k|_m \qquad \text{(C.16)}$$

and, using this summation, the following lemma can be adopted to bound the modulus matrix of $Q(z)^{-1}$ (i.e., $|Q(z)^{-1}|_m$).

Lemma C.12. *[98] In $|z| \geq 1$, the following inequality is satisfied $|Q(z)^{-1}|_m \leq \Sigma_{R_k}$.* ∎

C.3.3 The Stability of Interval Polynomial Matrices

In the remainder of this section, a stability condition for interval polynomial matrices is developed. For clarity of notation, from this point forward the superscript o is used to denote the nominal value of any variable or parameter. In particular:

Definition C.13. A^o denotes the nominal matrix of A^I:

$$A^o = \left[a_{ij}^o : a_{ij}^o = \frac{\underline{a_{ij}} + \overline{a_{ij}}}{2}\right].$$

∎

Likewise, $P(z)$, $S(z)$, and $Q(z)$ defined in (C.3) and (C.4) are now denoted by $P^o(z)$, $S^o(z)$, and $Q^o(z)$, respectively.

Now, let us add interval radius matrices ΔA_i to $S^o(z)$ to get the interval polynomial matrices $S^I(s)$ as follows:

$$S^I(z) = A_1^o + \Delta A_1 + (A_2^o + \Delta A_2)z^{-1} + \cdots + (A_n^o + \Delta A_n)z^{-n+1},$$

from which the interval coefficient matrices are defined element-wise as $A_i^o - |\Delta A_i|_m \leq A_i^I \leq A_i^o + |\Delta A_i|_m$. We also define:

$$Q^I(z) = zI + S^o(z) + \Delta S(z), \tag{C.17}$$

where $\Delta S(z) = \Delta A_1 + \Delta A_2 z^{-1} + \cdots + \Delta A_n z^{-n+1}$ and define the summation of the modulus interval matrices $|\Delta A_k|_m$ as

$$\triangle_M = \sum_{k=1}^{n} |\Delta A_k|_m. \tag{C.18}$$

In fact, in (C.16), it is not possible to estimate Σ_{R_k}, but if $Q^o(z)$ is stable, then Σ_{R_k} is bounded from (C.15). Let us suppose that the upper boundary of Σ_{R_k} is known as Σ^*. Then, we can find a condition for robustly stability of $Q^I(z)$. For this purpose, first, let us rewrite $\det[Q^I(z)]$, using $Q^I(z) = Q^o(z) + \Delta S(z)$, to be

$$\det[Q^o(z) + \Delta S(z)] = \det\left[Q^o(z)\left(I + (Q^o(z))^{-1}\Delta S(z)\right)\right]$$
$$= \det[Q^o(z)]\det\left[I + (Q^o(z))^{-1}\Delta S(z)\right]$$
$$= \det[Q^o(z)]\det\left[I + (Q^o(z))^{-1}\Delta S(z)\right]. \tag{C.19}$$

Then, based on Lemma C.8 and with the assumption of stable $Q^o(z)$, we have $\det[Q^o(z)] \neq 0$ for $|z| \geq 1$. Also, if $\rho\left((Q^o(z))^{-1}\Delta S(z)\right) < 1$, then, from Lemma C.4, we can say that

$$\det \left[I + (Q^o(z))^{-1} \, \Delta S(z) \right] \neq 0.$$

Thus, since $\det [Q^o(z)] \neq 0$ at $|z| \geq 1$, if

$$\rho \left((Q^o(z))^{-1} \, \Delta S(z) \right) < 1,$$

then the interval polynomial matrix, $Q^I(z)$, is stable. Therefore, we can conclude that $Q^I(z)$ is stable if $\rho \left((Q^o(z))^{-1} \, \Delta S(z) \right) < 1$. Next, let us investigate $\rho \left((Q^o(z))^{-1} \, \Delta S(z) \right) < 1$. Using Lemma C.6 and Lemma C.12, the following inequalities are true:

$$\rho \left((Q^o(z))^{-1} \, \Delta S(z) \right) \leq \rho \left(\left| (Q^o(z))^{-1} \right|_m |\Delta S(z)|_m \right) \leq \rho \left(\Sigma_{R_k} |\Delta S(z)|_m \right)$$
$$\leq \rho \left(\Sigma^* |\Delta S(z)|_m \right).$$

Furthermore, using $|z^{-k+1}|_m \leq 1$, when $|z| \geq 1$ and $k \geq 1$, the following relationships are true:

$$|\Delta S(z)|_m = \left| \sum_{k=1}^n \Delta A_k z^{-k+1} \right|_m \leq \sum_{k=1}^n |\Delta A_k|_m \left| z^{-k+1} \right|_m$$
$$\leq \sum_{k=1}^n |\Delta A_k|_m = \triangle_M,$$

where $\triangle_M$ was defined in (C.18). Then, by Lemma C.6, the following inequality is satisfied:

$$\rho \left((Q^o(z))^{-1} \, \Delta S(z) \right) \leq \rho (\Sigma^* \triangle_M).$$

Therefore, the following lemma can be developed.

Lemma C.14. *If $\rho(\Sigma^* \triangle_M) < 1$, then the interval polynomial matrix, $Q^I(z)$, is stable.* ∎

Using this result, the following theorem is finally developed for the robust stability of $P^I(z)$:

Theorem C.15. *If (i) $Q^o(z)$ is stable, (ii) $\rho(\Sigma^* \triangle_M) < 1$, and (iii) $\det(A_n^I) \neq 0$, then the interval polynomial matrix $P^I(z)$ is robustly stable.* ∎

Proof. From Lemma C.14, if $Q^o(z)$ is stable and $\rho(\Sigma^* \triangle_M) < 1$, then $Q^I(z)$ is robust stable. Also, from Theorem C.9, if $\det(A_n^I) \neq 0$, it is obvious that $P^I(z)$ is robustly stable. ∎

In the sequel, a method for analytically finding Σ^* is provided. Without notational confusion, $|\cdot|$ is the modulus matrix defined earlier. From (C.15),

$$|\overline{R}_{n+1}| + |\overline{R}_{n+2}| + \cdots + |\overline{R}_{n+p}| = |\mathcal{C}\overline{R}_n| + |\mathcal{C}\overline{R}_{n+1}| + \cdots + |\mathcal{C}\overline{R}_{n+p-1}|$$
$$= |\mathcal{C}\overline{R}_n| + |\mathcal{C}^2\overline{R}_n| + \cdots + |\mathcal{C}^p\overline{R}_n|$$
$$\leq (|\mathcal{C}| + |\mathcal{C}^2| + \cdots + |\mathcal{C}^p|) |\overline{R}_n|, \qquad \text{(C.20)}$$

where we used the inequality $|AB| \leq |A||B|$. Here, if $\mathcal{C}$ is diagonalizable as $\mathcal{C} = \mathcal{X}\Lambda\mathcal{X}^{-1}$, we then change (C.20) as follows:

$$(|\mathcal{C}| + |\mathcal{C}^2| + \cdots + |\mathcal{C}^p|) |\overline{R}_n| = (|\mathcal{X}\Lambda\mathcal{X}^{-1}| + |\mathcal{X}\Lambda^2\mathcal{X}^{-1}| + \cdots + |\mathcal{X}\Lambda^p\mathcal{X}^{-1}|) |\overline{R}_n|$$
$$\leq |\mathcal{X}| (|\Lambda| + |\Lambda^2| + \cdots + |\Lambda^p|) |\mathcal{X}^{-1}||\overline{R}_n|, \qquad \text{(C.21)}$$

which yields the following general formula:

$$\sum_{i=1}^{p} |\overline{R}_{n+i}| \leq |\mathcal{X}| \left(\sum_{i=1}^{p} |\Lambda^i| \right) |\mathcal{X}^{-1}||\overline{R}_n|. \qquad \text{(C.22)}$$

Now, taking $p \to \infty$, if $Q^o(z)$ is stable, then we have

$$\lim_{p\to\infty} \sum_{i=1}^{p} |\overline{R}_{n+i}| = |\mathcal{X}| \lim_{p\to\infty} \left(\sum_{i=1}^{p} |\Lambda^i| \right) |\mathcal{X}^{-1}||\overline{R}_n|$$
$$\leq |\mathcal{X}|\mathrm{diag}\left(\frac{|\lambda_l|}{1 - |\lambda_l|} \right) |\mathcal{X}^{-1}||\overline{R}_n|, \qquad \text{(C.23)}$$

where $\mathrm{diag}(\cdot)$ is a diagonal matrix composed of diagonal terms $(\cdot)$. Since $\mathcal{X}$, λ_l, $\mathcal{X}^{-1}$, and $\overline{R}_n$ are known, we can estimate the boundary of $\lim_{p\to\infty} \sum_{i=1}^{p} |\overline{R}_{n+i}|$. Writing $\mathcal{T} := |\mathcal{X}|\mathrm{diag}\left(\frac{|\lambda_l|}{1-|\lambda_l|} \right) |\mathcal{X}^{-1}|$ and $\mathcal{F}_n := \mathcal{T}|\overline{R}_n|$, and taking the first m rows of $\mathcal{F}_n$, which is denoted as $\mathcal{D}_n$ (i.e., $\mathcal{D}_n := \mathcal{F}_n(1:m, 1:m)$), we have the following inequality:

$$\Sigma_{R_k} \leq \sum_{i=1}^{n} |R_i| + \mathcal{D}_n. \qquad \text{(C.24)}$$

Therefore, since $\sum_{i=1}^{n} |R_i|$ and $\mathcal{D}_n$ are calculated, we can analytically estimate the upper boundary of Σ_{R_k}. However, (C.24) could be conservative. To accurately estimate the upper boundary of Σ_{R_k}, by introducing an operator $\mathcal{F}_q := \mathcal{T}|\overline{R}_q|$, where $q \gg n$, and writing $\mathcal{D}_q := \mathcal{F}_q(1:m, 1:m)$, we have a more accurate upper boundary of Σ_{R_k} given as

$$\Sigma_{R_k} \leq \sum_{i=1}^{q} |R_i| + \mathcal{D}_q := \Sigma^*. \qquad \text{(C.25)}$$

This argument about an accurate upper boundary of Σ_{R_k} is summarized in the following theorem:

Theorem C.16. *If $Q^o(z)$ is stable, in (C.25), as $q \to \infty$, $\mathcal{D}_q \to 0$; hence $\Sigma^* \to \Sigma_{R_k}$.*   ∎

Proof. Since $\mathcal{T}$ is fixed, from (C.15), $|\overline{R}_q| \to 0$ as $q \to \infty$ if and only if $\rho(\mathcal{C}) < 1$. Therefore, since $|\overline{R}_q| = 0$ if and only if $\overline{R}_q = 0$, the proof is immediate. ∎

Theorem C.16 shows that an accurate upper boundary of Σ_{R_k} (i.e., Σ^*) can be estimated by taking a very large q in (C.25).

C.4 Illustrative Examples

In this section, we test the conservatism of the suggested algorithm. From the existing literature, however, it is difficult to find a benchmark example for the robust stability of discrete polynomial matrices. Most examples are continuous cases [243, 232, 476, 191, 372, 481], except examples 3.5 and 4.3 of [372]. However, example 3.5 of [372] is a special case that allows the analytical calculation of $P(\lambda)^{-1}$. But, in general it is very difficult to calculate $P(\lambda)^{-1}$ analytically. Thus, in this appendix, for comparison purposes we use example 4.3 of [372], which uses a numerical range for $\|P(\lambda)^{-1}\|_2$.

C.4.1 Example 1

The main disadvantage of the suggested method in [372] is that it calculates $\|P(\lambda)^{-1}\|_2$ analytically. Although [372] provides a method for this, the method is quite complicated and the result could be very conservative (see Theorem 4.1 of [372]). In example 4.3 of [372], the stability radius of a polynomial matrix was calculated under some conditions. Let us use the example of [372], given as

$$P^o(z) = Iz^3 + A_1^o z^2 + A_2^o z + A_3^o, \tag{C.26}$$

where the coefficient matrices are Hermitian and satisfy the conditions $0 \leq \lambda_{min}(A_1^o) \leq \lambda_{max}(A_1^o) \leq 1/3$, $-1/9 \leq \lambda_{min}(A_2^o) \leq \lambda_{max}(A_2^o) \leq 1/9$, and $-1/27 \leq \lambda_{min}(A_3^o) \leq \lambda_{max}(A_3^o) \leq 1/9$. Since [372] does not provide the coefficient matrices, we selected the following matrices, which satisfy the conditions of example 4.3 of [372]:

$$A_1^o = \begin{bmatrix} 0 & 0 & 0 \\ 0 & 0.05 & 0 \\ 0 & 0 & 0.3333 \end{bmatrix} ; \ A_2^o = \begin{bmatrix} -0.1111 & 0 & 0 \\ 0 & 0.01 & 0 \\ 0 & 0 & 0.1111 \end{bmatrix}$$

$$A_3^o = \begin{bmatrix} -0.0370 & 0 & 0 \\ 0 & 0.05 & 0 \\ 0 & 0 & 0.1111 \end{bmatrix}.$$

In [372], the analytical perturbation radius of $P^o(z)$ is calculated as 0.0141, i.e., $\|[0_{3\times3} \ \Delta_1 \ \Delta_2 \ \Delta_3]\|_2$ could be 0.0141. Thus, since $\|[0_{3\times3} \ \Delta_1 \ \Delta_2 \ \Delta_3]\|_2 \leq 0.0141$, the polynomial matrix system is robustly stable. Let us use our method to compute this analytical stability radius. From the companion form $\mathcal{C}$, we find that the nominal system is stable and has non-repeated eigenvalues. From Theorem C.16 and (C.25), selecting $q = 100$, we found that

$$\Sigma^* = \sum_{i=1}^{q} |R_i| + \mathcal{D}_q = \begin{bmatrix} 1.1739 & 0 & 0 \\ 0 & 1.1152 & 0 \\ 0 & 0 & 1.5 \end{bmatrix}.$$

Also, for the uncertainty, we provided 10 percent intervals to A_1^o, A_2^o, and A_3^o, from which $\|[0_{3\times3} \ \ \Delta_1 \ \ \Delta_2 \ \ \Delta_3]\|_2 = 0.0369$. From (C.18), we calculated

$$\triangle_M = \begin{bmatrix} 0.0148 & 0 & 0 \\ 0 & 0.011 & 0 \\ 0 & 0 & 0.0556 \end{bmatrix}$$

Finally, using the calculated $\sum_{i=1}^{q} |R_i| + \mathcal{D}_q$ and $\triangle_M$, we found that $\rho(\Sigma^* \triangle_M) = 0.0833$; hence the interval polynomial system is robustly stable with 10 percent uncertainty, which cannot be concluded in [372].

C.4.2 Example 2

Let us first consider the following general non-symmetric polynomial matrix:

$$P^o(z) = I_{2\times2} z^3 + A_1^o z^2 + A_2^o z + A_3^o, \tag{C.27}$$

where the coefficient matrices are given as

$$A_1^o = \begin{bmatrix} 0.4 & -0.3 \\ 0.4 & 0.1 \end{bmatrix}; \ A_2^o = \begin{bmatrix} 0.3 & 0.3 \\ 0.4 & -0.5 \end{bmatrix}$$

$$A_3^o = \begin{bmatrix} 0.0 & 0.5 \\ -0.1 & 0.15 \end{bmatrix}.$$

From the corresponding $\mathcal{C}$ matrix, the eigenvalues are calculated as $-0.0535 + 0.8125i$; $-0.0535 - 0.8125i$; -0.8696; -0.5086; 0.7612; 0.2240. Thus, the nominal system is stable. Also, since $\det(A_3^I)$ is not zero, the suggested method can be used. It is assumed that there exists element-wise interval uncertainty in the nominal matrices given as

$$\Delta A_1 = \begin{bmatrix} 0.0432 & 0.0324 \\ 0.0432 & 0.0108 \end{bmatrix}; \ \Delta A_2 = \begin{bmatrix} 0.0324 & 0.0324 \\ 0.0432 & 0.0540 \end{bmatrix}$$

$$\Delta A_3 = \begin{bmatrix} 0.0 & 0.0540 \\ 0.0108 & 0.0162 \end{bmatrix}.$$

To apply Theorem C.16, we selected $q = 50$, from which we found

$$\Sigma^* = \sum_{i=1}^{50} |R_i| + \mathcal{D}_{50} = \begin{bmatrix} 3.1502 & 1.8140 \\ 1.5848 & 4.1860 \end{bmatrix}.$$

Using these matrices, we found $\rho(\Sigma^* \triangle_M) = 0.9979$ which shows that the system is robustly stable (though almost marginally stable). For comparison purposes, we used Theorem 3.1 of [372]. It is, however, difficult to find the infimum of $\dfrac{1}{\sqrt{\sum_{k \in J} |\lambda|^{2k} \|P(\lambda)^{-1}\|_2}}$ for all $\lambda \in \partial \Omega$, which is a required computation (for notation, refer to [372]). Hence we performed random simulation tests and found that $\lambda = -1$ is the best (this is not an analytical solution, instead we did 2000 random tests to find the best λ). Using $\lambda = -1$, we calculated $\dfrac{1}{\sqrt{\sum_{k \in J} |\lambda|^{2k} \|P(\lambda)^{-1}\|_2}} = 0.1175$, and calculated $\|[\triangle_0 \ \ \triangle_1 \ \ \triangle_2 \ \ \triangle_3]\| = 0.1174$ which is also almost marginally stable. Hence, from this test, we found that when the exact minimum of $\dfrac{1}{\sqrt{\sum_{k \in J} |\lambda|^{2k} \|P(\lambda)^{-1}\|_2}}$ of Theorem 3.1 of [372] is found, the stability index of [372] is almost equal to the stability index of our method. However, as commented in [372], it is tough to find the exact minimum of $\dfrac{1}{\sqrt{\sum_{k \in J} |\lambda|^{2k} \|P(\lambda)^{-1}\|_2}}$ (so they used a numerical range for the approximation, but the result is quite conservative as shown in Example 1 above).

C.5 Summary

In this appendix, a new method for checking the Schur stability of interval polynomial matrices has been suggested and illustrated. The proposed method checks the stability in a simple manner and the derivation process of the method has a good analytical basis. From comparison with an existing method, it was found that the suggested method is less conservative, as demonstrated in Example 1 above, and computationally very simple. Furthermore, we have found that our method provides almost the same stability condition as [372] when the exact minimum of Theorem 3.1 of [372] is found.[3] This implies that the analytical solution proposed in this appendix can be very useful.

---

[3] However, actually it is very tough to find this minimum. So, [372] developed Theorem 4.1, which results in a conservative stability radius.

D

Power of an Interval Matrix

Interval computation techniques are popularly used for robust stability analysis of uncertain systems described in terms of interval parameters and interval matrices. In the past two decades, a great amount of research effort has been devoted to the analysis of interval matrices. However, there is no available result for determining the impulse response bounds of discrete-time, linear, time-invariant systems with interval uncertainty in their state-space description. Indeed, in control engineering [245, 52] and in signal processing [297], the impulse response plays an important role. As shown in [553], the interval impulse response could be effectively used for robust controller design. However for all these works, if there exist model uncertainties in a system's state-space model, the uncertain ranges of the impulse response should be carefully estimated.

Note that if the interval uncertainty in the state-space model is known then the key problem of determining the impulse response of such an uncertain system is to find the power of an interval matrix. In this appendix we will show how the power of an interval matrix can be computed. This result can then be used for a number of important problems. For our purposes, for example, it can be used for designing the learning gain matrix in ILC problems (although, the method presented here requires much more computation compared with the eigen-decomposition method developed in Chapter 6). Beyond ILC, in robust control the power of an interval matrix can be effectively used in the analysis of controllability, observability, or impulse response for uncertain interval systems. However, very limited effort has been devoted to calculating the boundaries of the power of an interval matrix. Some existing results can be found in [290, 202, 460, 172] where the convergence problem of the powers of an interval matrix was studied and it was proved that the power of an interval matrix converges to zero if the maximum spectral radius of the interval matrix is less than 1. However, the question of the boundaries of the power of an interval matrix at a specified order has not been fully addressed (though, some useful analysis of the power of an interval matrix at a specified order can be found in [244], where it was concluded that computing the boundaries of the power of an interval matrix is an NP-hard problem).

D.1 Sensitivity Transfer Method

In this section, a new method is developed for the calculation of the power of an interval matrix. This method first computes the sensitivity of the perturbation of the nominal A matrix and then applies this sensitivity to the power of the matrix A^k. The set of the power of the interval matrix can be written as

$$\mathcal{A}^k = \{P \mid P = \underbrace{AAA \cdots A}_{k}, \ A \in A^I\}. \tag{D.1}$$

Then, from the relationship $A^k = \underbrace{AA \cdots A}_{k}$, we can have

$$\frac{\partial A^k}{\partial a_{ij}} = \frac{\partial A}{\partial a_{ij}} (\underbrace{A \cdots A}_{k-1}) + A \frac{\partial A}{\partial a_{ij}} (\underbrace{A \cdots A}_{k-2}) + \cdots + (\underbrace{A \cdots A}_{k-1}) \frac{\partial A}{\partial a_{ij}}. \tag{D.2}$$

Here, by observing that $\frac{\partial A}{\partial a_{ij}} = I_{ij}$ where I_{ij} is a matrix whose i^{th} row and j^{th} column element is 1 and the other elements are all zeroes, we have

$$\frac{\partial A^k}{\partial a_{ij}} = I_{ij} (\underbrace{A \cdots A}_{k-1}) + A I_{ij} (\underbrace{A \cdots A}_{k-2}) + \cdots + (\underbrace{A \cdots A}_{k-1}) I_{ij}. \tag{D.3}$$

Thus, we have the perturbed sensitivity (∂A^k) of A^k by the uncertain change (∂a_{ij}) of a_{ij} such as

$$\partial A^k = \partial a_{ij} \left(I_{ij} (\underbrace{A \cdots A}_{k-1}) + A I_{ij} (\underbrace{A \cdots A}_{k-2} + \cdots + (\underbrace{A \cdots A}_{k-1}) I_{ij} \right). \tag{D.4}$$

For convenience, let us use the following notation:

$$\prod_{ij} := \left(I_{ij} (\underbrace{A \cdots A}_{k-1}) + A I_{ij} (\underbrace{A \cdots A}_{k-2}) + (\underbrace{A \cdots A}_{2}) I_{ij} (\underbrace{A \cdots A}_{k-3} + \cdots + (\underbrace{A \cdots A}_{k-1}) I_{ij} \right)$$

which simplifies (D.4) to $\partial A^k = \partial a_{ij} \prod_{ij}$. Hence, we find that when there is a perturbation amount of ∂a_{ij} in a_{ij}, there is a perturbation effect on A^k by the amount of ∂A^k that is related to the sensitivity transfer matrix $\prod_{ij}$. Here, noticing that each element of A perturbs A^k, we develop a method for bounding the uncertainty of A^k. Using the notation $P \in \mathcal{P}^1 = \mathcal{A}^k = [\underline{P}, \overline{P}]$, we make the following proposition:

Proposition D.1. *Given the order of power k, the upper and lower boundaries associated with the elements of P occur at the power of one of the vertex matrices A^v.* ∎

Proof. Let us pick arbitrary i_1 and j_1, and fix all a_{pq}, where $p, q = 1, \ldots, n$, and $p \neq i_1$ or $q \neq j_1$, to specified values $a_{pq} = a_{pq}^* \in [\overline{a_{pq}}, \underline{a_{pq}}]$. Then, from $\partial A^k = \partial a_{i_1 j_1} \prod_{i_1 j_1}$, the kth row and l^{th} column element of ∂A^k are determined by $\partial a_{i_1 j_1}$ and $\left(\prod_{i_1 j_1} \right)_{kl}$. Noticing that $\partial a_{i_1 j_1} = [-\triangle a_{i_1 j_1}, \triangle a_{i_1 j_1}]$, the positive (negative) maximum of ∂A^k occurs at $\triangle a_{i_1 j_1}$ $(-\triangle a_{i_1 j_1})$ if $\left(\prod_{i_1 j_1} \right)_{kl}$ is positive. Otherwise, the positive (negative) maximum of ∂A^k occurs at $-\triangle a_{i_1 j_1}$ $(\triangle a_{i_1 j_1})$. However, the sign of $\left(\prod_{i_1 j_1} \right)_{kl}$ is not determined. Hence, for arbitrary fixed i_1 and j_1, we can conclude that the positive (negative) maximum of the kth row and l^{th} column element of ∂A^k occurs at one of the vertex points of $a_{i_1 j_1}^I = [a_{i_1 j_1}^o - \triangle a_{i_1 j_1}, a_{i_1 j_1}^o + \triangle a_{i_1 j_1}]$. Now let us pick another arbitrary i_2 and j_2. Then by the same reasoning given above, the positive (negative) maximum of the kth row and l^{th} column element of ∂A^k occurs at one of vertex points of $a_{i_2 j_2}^I = [a_{i_2 j_2}^o - \triangle a_{i_2 j_2}, a_{i_2 j_2}^o + \triangle a_{i_2 j_2}]$, but $a_{i_1 j_1} \in \{\underline{a_{i_1 j_1}}, \overline{a_{i_1 j_1}}\}$. Finally, since we can repeat the above discussion for all a_{pq}, the positive (negative) maximum of the kth row and l^{th} column element of ∂A^k occurs at the power of one of vertex matrices. ∎

Proposition D.1 shows that the lower and upper boundaries of the power of an interval matrix can be found by checking all the vertex matrices. It is important to highlight that Proposition D.1 uses the finite vertex matrix set to find the boundary of the power of interval matrix set. However, from $\partial A^k = \partial a_{ij} \prod_{ij}$, it is required to check all the vertex matrices of A^I to find the maximal positive or negative perturbation of elements of A^k, i.e., $(A^k)_{ij}$, $\forall A \in A^I$. Hence, the computational amount could be huge. That is, in order to find the maximum and minimum of A^k, where $A \in A^I$, we have to check 2^{n^2} vertex matrices, where n is the size of the square A matrix, for each element of A^k. Thus, the total computational amount is $2^{n^2} \times 2^n = 2^{n^2 + n}$. In the sequel, we will show that under some conditions, we do not need to check all the vertex matrices. Instead, only some specified vertex matrices need to be used for the calculation of the power of interval matrix. However, even without this result, although the computational effort to check all the vertex matrices may be high, because the impulse response of an LTI system is generally used for design purposes in an off-line manner, Proposition D.1 is still useful.

To see how to reduce the computational load, let us define the center matrix of P and radius matrix of P as $P^c = \frac{P + \overline{P}}{2}$; $P^r = \frac{\overline{P} - \underline{P}}{2}$. Then the following result can be derived:

Proposition D.2. *If the sign of the kth row and l^{th} column element of the sensitivity transfer matrix $\prod_{ij}$ does not change by ∂a_{ij}, the maximum positive* *and negative perturbations of the kth row and l^{th} column element of $\mathcal{A}^k$, occur at the power of the following vertex matrices of A^I, respectively,*

$$(A^v)|_{kl}^+ = \left\{ A^v \mid A^v = [(A^o)_{ij} + s_{kl}^{ij} \triangle a_{ij}, i, j = 1, \dots, n] \right\} \qquad \text{(D.5)}$$

$$(A^v)|_{kl}^- = \left\{ A^v \mid A^v = [(A^o)_{ij} - s_{kl}^{ij} \triangle a_{ij}, i, j = 1, \dots, n] \right\}, \qquad \text{(D.6)}$$

where $s_{kl}^{ij} = \text{sign} \left(\Pi_{ij} \right)_{kl}$. ∎

Proof. From $\partial A^k = \partial a_{ij} \Pi_{ij}$, the positive (negative) maximum of ∂A^k occurs at $\triangle a_{ij}$ $(-\triangle a_{ij})$ if $\left(\Pi_{ij} \right)_{kl}$ is positive. Otherwise, the positive (negative) maximum of ∂A^k occurs at $-\triangle a_{ij}$ $(\triangle a_{ij})$. This implies that the positive (negative) maximum disturbance of $(A^k)_{kl}$ occurs at $(A^o)_{ij} + s_{kl}^{ij} \triangle a_{ij}$ $((A^o)_{ij} - s_{kl}^{ij} \triangle a_{ij})$. ∎

Remark D.3. Proposition D.2 was developed based on an assumption that the signs of Π_{ij} do not change. In Section D.3, some sufficient conditions are established which can be used for checking the sign variation. However, it should be noted that if the sufficient conditions given in the appendix are not satisfied, Proposition D.1 should be used. Hence, Proposition D.1 and Proposition D.2, having their own advantages and disadvantages, complement each other. Therefore, the two procedures based on the sensitivity transfer idea presented in this appendix are practically valuable in the interval model conversion problem.

Remark D.4. In Proposition D.1 and Proposition D.2, we considered general non-symmetric square interval matrices. However, we can extend these results to symmetric interval matrices. This work is direct by repeating (D.2), (D.3), and (D.4).

In this section, a new method called the sensitivity transfer method was developed to overcome the conservatism of the method suggested in Chapter 6 (called eigenpair-decomposition method). However, the computational effort of the method given in Chapter 6 is significantly less than the method developed in this appendix. Using either method, however, once the boundary of the power of an interval matrix is found then it is straightforward to find the boundaries of the impulse response. In other words, the boundaries of h_k^I, i.e., $[\underline{h}_k, \overline{h}_k]$, can be simply calculated by matrix multiplication $CP^{k-1}B$, where $P^{k-1} \in \mathcal{A}^{k-1}$, because the upper and lower boundary matrices of $\mathcal{A}^{k-1}$ have been estimated. In the next section, the effectiveness of the suggested method is illustrated through numerical examples.

D.2 Illustrative Examples

To verify the usefulness of the new method, Monte-Carlo-type random tests are performed. The results obtained from the random tests are considered as the "true" range of the impulse response of the interval LTI system, for comparison to the bounds computed by the suggested methods.

D.2.1 Example 1

Consider the following uncertain discrete-time LTI interval system:

$$\begin{aligned} x(t+1) &= Ax(t) + Bu(t) \\ y(t) &= Cx(t), \end{aligned} \tag{D.7}$$

where $B = [2, 0.5]^T$; $C = [1, 0]$ and the following two different nominal A matrices are tested:

- Case 1 (symmetric and unstable): $a_{11} = -1.05$; $a_{12} = 0.55$; $a_{21} = 0.55$; $a_{22} = 0.85$
- Case 2 (non-symmetric and stable): $a_{11} = 0.75$; $a_{12} = -0.40$; $a_{21} = 0.25$; $a_{22} = 0.55$.

In Case 1, the nominal eigenvalues are -1.1977 and 0.9977. Thus, the nominal system is initially unstable. In Case 2, the nominal eigenvalues are $0.6500 + 0.3000i$ and $0.6500 - 0.3000i$. Hence, the nominal system is stable. In both cases, it is supposed that there is 10 percent interval model uncertainty in the A matrix parameters. Thus, in Case 1, A^I is

$$A^I = \begin{bmatrix} [-1.155, -0.945] & [0.495, 0.605] \\ [0.495, 0.605] & [0.765, 0.935] \end{bmatrix}$$

and in Case 2, $\mathcal{A}^I$ is

$$A^I = \begin{bmatrix} [0.675, 0.825] & [-0.44, -0.36] \\ [0.225, 0.275] & [0.495, 0.605] \end{bmatrix}.$$

Figure D.1 shows the test result of Case 1. Since the system is unstable, the impulse responses diverge as k increases. In this figure, four different test results are shown: the ×-dot dashed lines are the upper/lower boundaries computed from Intlab [179]; the ○-dashed lines show the upper/lower boundaries computed from the eigenpair-decomposition method; the ◇-solid lines represent the upper/lower boundaries computed from the sensitivity transfer method; and the thick solid vertical bars represent the range obtained from the random test results. Clearly, the sensitivity transfer method accurately bounds the upper/lower boundaries of the impulse responses, while even if the eigenpair-decomposition method is better than Intlab, it is much more conservative than the sensitivity transfer method. Figure D.2 shows the test results of Case 2. From this figure, it is also seen that the sensitivity transfer method accurately bounds the upper/lower boundaries of the impulse responses. In the early phase, Intlab performs better than the eigenpair-decomposition-based method. However, as k increases, the eigenpair-decomposition method performs better than Intlab. Note that for the sensitivity transfer method, we used Proposition D.2 based on Lemma D.5 and Lemma D.6. However, the condition of Lemma D.6 was satisfied, for all k, l, i, j in Proposition D.2, only for the power $k = 1, \ldots, 4$ of Case 1 and the condition of Lemma D.5 was satisfied, for all k, l, i, j in Proposition D.2, only for the power $k = 1, \ldots, 4$ of Case 2.[1] Hence, for the higher-order power of the interval matrix, we used Proposition D.1. Now, from Figure D.1 and Figure D.2, it is clear that the sensitivity transfer method suggested in this appendix very accurately bounds the impulse responses of the uncertain interval system in both stable and unstable systems. However, the computational amount is huge. Thus we can say that the eigenpair-decomposition method and the sensitivity transfer method complement each other.

D.2.2 Example 2

For the usefulness of Proposition D.2 and Lemma D.5, let us consider the following nominal matrix, which was created using MATLAB® commands *rand* and *sign*:

$$A^o = \begin{bmatrix} 1 & -1 & -1 & -1 & 1 \\ -1 & 1 & -1 & 1 & -1 \\ 1 & -1 & -1 & 1 & -1 \\ -1 & 1 & 1 & 1 & 1 \\ 1 & -1 & 1 & 1 & 1 \end{bmatrix}$$

Let us suppose that there exist ± 0.001 interval uncertainties in all elements, and that we want to find exact upper and lower boundaries of $A^5, A \in A^I$. If we use Proposition D.1, we need to check $2^{5^2} = 2^{25}$ vertex matrices. Indeed, in this case, the computational time could be huge. However, from Lemma D.5, we find that the signs of all elements of $\prod_{ij}$ do not change for all i, j. Hence, for the upper and lower boundary matrices of A^5, it is enough to use 25 vertex matrices. From these vertex matrices, we calculate the upper boundary and lower boundary matrices of $A^5, A \in A^I$ to be

$$\overline{P} = \begin{bmatrix} 125.7868 & -124.2167 & -0.7030 & -68.4220 & 69.5497 \\ -130.1691 & 131.8351 & -0.6711 & 91.6420 & -90.3863 \\ 29.2777 & -28.6957 & -0.8434 & -4.7393 & 5.2731 \\ -62.3879 & 63.6182 & 11.2053 & 55.4559 & -22.5556 \\ 97.7063 & -96.2963 & 11.2414 & -40.4952 & 73.4820 \end{bmatrix}$$

---

[1] From numerous numerical tests, we have found that Lemma D.5 and Lemma D.6 are particularly effective for a stable system and a lower-order impulse response. Also it is important to emphasize that Proposition D.2 does not require that Lemma D.5 and Lemma D.6 should be satisfied for all k, l, i, j. Instead, Proposition D.2 shows that if Lemma D.5 and Lemma D.6 hold for part of k, l, i, j, the corresponding elements of the power of an interval matrix can be estimated from the particular intervals of A^I. In such a case, the computational effort could be further reduced.

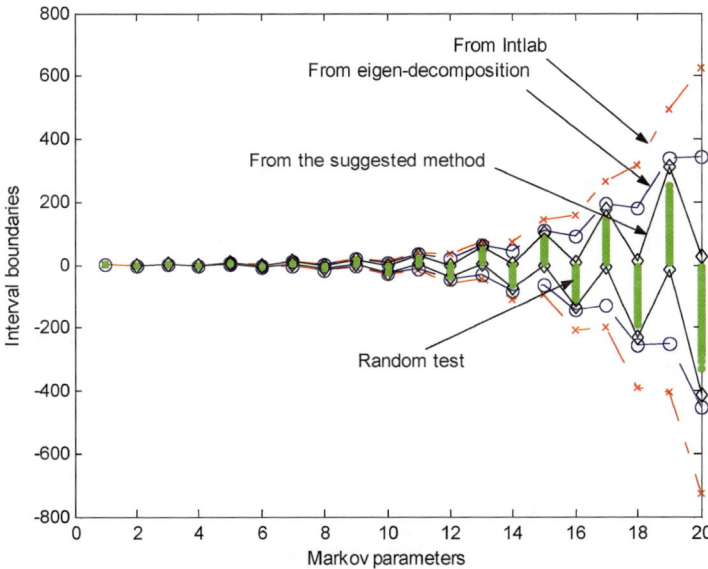

Fig. D.1. Impulse responses of Case 1. Plots are from `Intlab`, from the eigenpair-decomposition method, and from the suggested sensitivity transfer method. The vertical thick bars are the random test results.

$$P = \begin{bmatrix} 124.2167 & -125.7868 & -1.2970 & -69.5800 & 68.4517 \\ -131.8351 & 130.1691 & -1.3291 & 90.3600 & -91.6163 \\ 28.7237 & -29.3057 & -1.1574 & -5.2613 & 4.7271 \\ -63.6139 & 62.3842 & 10.7953 & 54.5459 & -23.4456 \\ 96.2963 & -97.7063 & 10.7594 & -41.5052 & 72.5200 \end{bmatrix}.$$

D.3 Condition for Proposition D.2

In this section we provide sufficient conditions for Proposition D.2. Let us write the sensitivity transfer matrix $\prod_{ij}$ as

$$\prod_{ij} = \sum_{p=1}^{k} A^{p-1} I_{ij} A^{k-p}, \tag{D.8}$$

where $A \in A^I$. For convenience, let us write A as $A = A^o + \Delta$ where $\Delta \in \Delta A^I$. Then, using $A^k = (A^o + \Delta)^k$, and writing $\mathcal{O}^k := (A^o + \Delta)^k - (A^o)^k$, we obtain:

$$\prod_{ij} = \sum_{p=1}^{k} (A^o + \Delta)^{p-1} I_{ij} (A^o + \Delta)^{k-p}$$

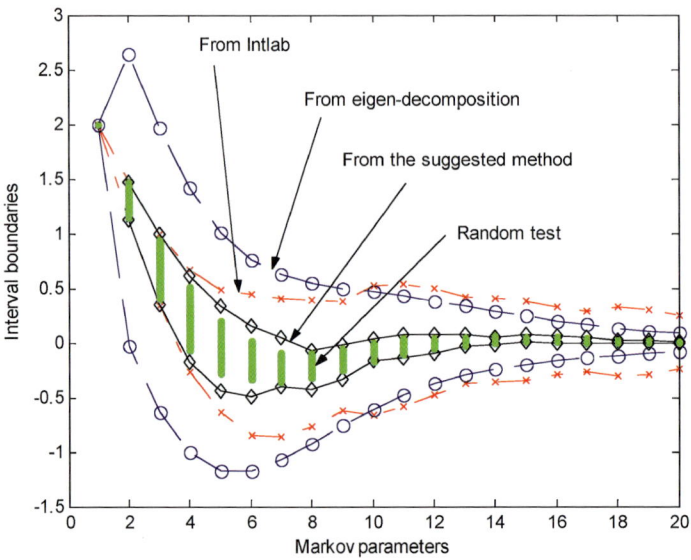

Fig. D.2. Impulse responses of Case 2. Plots are from Intlab, from the eigenpair decomposition method, and from the suggested sensitivity transfer method. The vertical thick bars are the random test results.

$$= \sum_{p=1}^{k} \left[\mathcal{O}^{p-1} + (A^o)^{p-1} \right] I_{ij} \left[\mathcal{O}^{k-p} + (A^o)^{k-p} \right]. \qquad (D.9)$$

Then, rearranging (D.9) yields

$$\prod_{ij} - \sum_{p=1}^{k} (A^o)^{p-1} I_{ij} (A^o)^{k-p} = \sum_{p=1}^{k} \{ \mathcal{O}^{p-1} I_{ij} \mathcal{O}^{k-p} + \mathcal{O}^{p-1} I_{ij} (A^o)^{k-p} \\ + (A^o)^{p-1} I_{ij} \mathcal{O}^{k-p} \}. \qquad (D.10)$$

Defining the absolute matrix such as $|A| := [|a_{ij}|]$, we have

$$\left| \prod_{ij} - \sum_{p=1}^{k} (A^o)^{p-1} I_{ij} (A^o)^{k-p} \right| =$$

$$\left| \sum_{p=1}^{k} \{ \mathcal{O}^{p-1} I_{ij} \mathcal{O}^{k-p} + \mathcal{O}^{p-1} I_{ij} (A^o)^{k-p} + (A^o)^{p-1} I_{ij} \mathcal{O}^{k-p} \} \right|$$

$$\leq \sum_{p=1}^{k} \{ |\mathcal{O}^{p-1}| I_{ij} |\mathcal{O}^{k-p}| + |\mathcal{O}^{p-1}| I_{ij} |(A^o)^{k-p}| + |(A^o)^{p-1}| I_{ij} |\mathcal{O}^{k-p}| \}$$

$$\leq \sum_{p=1}^{k} \left\{ \left[(|A^o| + |\Delta|)^{p-1} - |A^o|^{p-1} \right] I_{ij} \left[(|A^o| + |\Delta|)^{k-p} - |A^o|^{k-p} \right] \right.$$

$$+ \left[(|A^o| + |\Delta|)^{p-1} - |A^o|^{p-1} \right] I_{ij} \left| (A^o)^{k-p} \right|$$

$$\left. + \left| (A^o)^{p-1} \right| I_{ij} \left[(|A^o| + |\Delta|)^{k-p} - |A^o|^{k-p} \right] \right\}, \tag{D.11}$$

where we used the inequality $|\mathcal{O}^k| \leq \left[|A^o| + |\Delta| \right]^k - |A^o|^k$, which can be derived after several algebraic manipulations. Now, defining $\Delta^* := \overline{A} - A^o = A^o - \underline{A}$ and using inequality $|\Delta| \leq \Delta^*$, we obtain

$$\left| \prod_{ij} - \sum_{p=1}^{k} (A^o)^{p-1} I_{ij} (A^o)^{k-p} \right| \leq$$

$$\sum_{p=1}^{k} \left\{ \left[(|A^o| + \Delta^*)^{p-1} - |A^o|^{p-1} \right] I_{ij} \left[(|A^o| + \Delta^*)^{k-p} - |A^o|^{k-p} \right] \right.$$

$$+ \left[(|A^o| + \Delta^*)^{p-1} - |A^o|^{p-1} \right] I_{ij} \left| (A^o)^{k-p} \right|$$

$$\left. + \left| (A^o)^{p-1} \right| I_{ij} \left[(|A^o| + \Delta^*)^{k-p} - |A^o|^{k-p} \right] \right\}. \tag{D.12}$$

Finally, denoting the right-hand side of (D.12) as R^* and writing

$$L := \left| \sum_{p=1}^{k} (A^o)^{p-1} I_{ij} (A^o)^{k-p} \right|,$$

we can make the following lemma.

Lemma D.5. *If* $L \geq R^*$ *element-wise, the signs of* $\prod_{ij}$ *do not change element-wise.* ∎

Proof. From

$$\left| \prod_{ij} - \sum_{p=1}^{k} (A^o)^{p-1} I_{ij} (A^o)^{k-p} \right| \leq R^* \leq L$$

$$= \left| \sum_{p=1}^{k} (A^o)^{p-1} I_{ij} (A^o)^{k-p} \right|,$$

we have the inequality:

$$\left| \prod_{ij} - \sum_{p=1}^{k} (A^o)^{p-1} I_{ij} (A^o)^{k-p} \right| \leq \left| \sum_{p=1}^{k} (A^o)^{p-1} I_{ij} (A^o)^{k-p} \right|.$$

Hence element-wise, if $\sum_{p=1}^{k} (A^o)^{p-1} I_{ij} (A^o)^{k-p} \geq 0$, then

$$0 \le \prod_{ij} \le 2 \left(\sum_{p=1}^{k} (A^o)^{p-1} I_{ij} (A^o)^{k-p} \right),$$

else if $\sum_{p=1}^{k} (A^o)^{p-1} I_{ij} (A^o)^{k-p} < 0$, then

$$-2 \left(\sum_{p=1}^{k} (A^o)^{p-1} I_{ij} (A^o)^{k-p} \right) \le \prod_{ij} < 0.$$

Therefore, the signs of $\prod_{ij}$ are the same to the signs of $\sum_{p=1}^{k} (A^o)^{p-1} I_{ij} (A^o)^{k-p}$. This completes the proof. ∎

When the commutative property $A^o \Delta = \Delta A^o$ holds, a less conservative condition can be derived. Note the commutative property is satisfied when A is a symmetric interval matrix, and the symmetric interval matrix system has been an important research topic as shown in [192, 389]. For this result, we use $(A^o + \Delta)^m = \sum_{u=0}^{m} {}_mC_u (A^o)^{m-u} \Delta^u$ where ${}_mC_u = \frac{m!}{u!(m-u)!}$. Now, from the following relationship:

$$\prod_{ij} = \sum_{p=1}^{k} \left[\sum_{u=0}^{p-1} {}_{p-1}C_u (A^o)^{p-1-u} \Delta^u \right] I_{ij} \left[\sum_{v=0}^{k-p} {}_{k-p}C_v (A^o)^{k-p-v} \Delta^v \right]$$

$$= \sum_{p=1}^{k} \left[(A^o)^{p-1} + \sum_{u=1}^{p-1} {}_{p-1}C_u (A^o)^{p-1-u} \Delta^u \right] I_{ij}$$

$$\times \left[(A^o)^{k-p} + \sum_{v=1}^{k-p} {}_{k-p}C_v (A^o)^{k-p-v} \Delta^v \right],$$

(D.13)

we have

$$\prod_{ij} - \sum_{p=1}^{k} (A^o)^{p-1} I_{ij} (A^o)^{k-p} = \sum_{p=1}^{k} \left\{ \left[(A^o)^{p-1} \right] I_{ij} \left[\sum_{v=1}^{k-p} {}_{k-p}C_v (A^o)^{k-p-v} \Delta^v \right] \right.$$

$$+ \left[\sum_{u=1}^{p-1} {}_{p-1}C_u (A^o)^{p-1-u} \Delta^u \right] I_{ij} \left[(A^o)^{k-p} \right]$$

$$+ \left[\sum_{u=1}^{p-1} {}_{p-1}C_u (A^o)^{p-1-u} \Delta^u \right] I_{ij}$$

$$\left. \times \left[\sum_{v=1}^{k-p} {}_{k-p}C_v (A^o)^{k-p-v} \Delta^v \right] \right\}$$

$$= \sum_{p=1}^{k} \sum_{v=1}^{k-p} {}_{k-p}C_v \left[(A^o)^{p-1} I_{ij} (A^o)^{k-p-v} \Delta^v \right]$$

$$+ \sum_{p=1}^{k} \sum_{u=1}^{p-1} {}_{p-1}C_u \left[(A^o)^{p-1-u} \Delta^u I_{ij} (A^o)^{k-p} \right]$$

$$+ \sum_{p=1}^{k} \sum_{u=1}^{p-1} \sum_{v=1}^{k-p} ({}_{p-1}C_u)({}_{k-p}C_v)$$

$$\times \left[(A^o)^{p-1-u} \Delta^u I_{ij} (A^o)^{k-p-v} \Delta^v \right]. \quad \text{(D.14)}$$

Using the commutative property (notice that I_{ij} is symmetric), we simplify (D.14) as

$$\prod_{ij} - \sum_{p=1}^{k} (A^o)^{p-1} I_{ij} (A^o)^{k-p} = \sum_{p=1}^{k} \sum_{v=1}^{k-p} {}_{k-p}C_v \left[I_{ij} (A^o)^{k-v-1} \Delta^v \right]$$

$$+ \sum_{p=1}^{k} \sum_{u=1}^{p-1} {}_{p-1}C_u \left[\Delta^u I_{ij} (A^o)^{k-u-1} \right]$$

$$+ \sum_{p=1}^{k} \sum_{u=1}^{p-1} \sum_{v=1}^{k-p} ({}_{p-1}C_u)({}_{k-p}C_v)$$

$$\left[\Delta^{u+v} I_{ij} (A^o)^{k-u-v-1} \right].$$

$$\text{(D.15)}$$

Hence, we obtain the following inequality:

$$\left| \prod_{ij} - \sum_{p=1}^{k} (A^o)^{p-1} I_{ij} (A^o)^{k-p} \right| \leq \sum_{p=1}^{k} \left\{ \sum_{v=1}^{k-p} {}_{k-p}C_v \left| I_{ij} (A^o)^{k-v-1} \right| (\Delta^*)^v \right.$$

$$+ \sum_{u=1}^{p-1} {}_{p-1}C_u (\Delta^*)^u \left| I_{ij} (A^o)^{k-u-1} \right|$$

$$+ \sum_{u=1}^{p-1} \sum_{v=1}^{k-p} ({}_{p-1}C_u)({}_{k-p}C_v)(\Delta^*)^{u+v}$$

$$\left. \times \left| I_{ij} (A^o)^{k-u-v-1} \right| \right\}.$$

$$\text{(D.16)}$$

Now, denoting the right-hand side of (D.16) as S^*, we can state the following lemma for the case of a symmetric interval matrix.

Lemma D.6. *For a symmetric interval matrix, if $L \geq S^*$ element-wise, the signs of $\prod_{ij}$ do not change.* ∎

The following remark is provided for some special cases.

- In Proposition D.2, in the case of $A > 0$ element-wise, for all $A \in A^I$, or $A < 0$ element-wise, for all $A \in A^I$, the signs of $\prod_{ij}$ do not change.

- When A^I is a symmetric interval matrix and it satisfies the property that, regardless of the magnitude of the elements of A^I, the sign of (AA) = the sign of (A) for all $A \in A^I$, then the signs of $\prod_{ij}$ do not change.

D.4 Summary

Computing the boundaries of the power of an interval matrix is a hard problem, if not NP-hard. In this appendix, we provided a solution for computing the bounds of the power of an interval matrix using the idea of sensitivity transfer. Through rigorous analysis, we are able to show that the exact boundaries of the power of an interval matrix can be found from vertex matrices.

Furthermore, in some special cases when the considered interval matrix has some structural constraints, the exact boundaries of the power of an interval matrix can be calculated from a set of selected vertex matrices. Numerical examples were presented to illustrate the proposed algorithm.

We believe that the results of this appendix can be widely used in solving many robust control problems such as the robust stability, robust controllability/observability, and others.

References

[1] C. T. Abdallah, V. S. Soualian, and E. Schamiloglu. Toward "smart tubes" using iterative learning control. *IEEE Trans. on Plasma Science*, 26(3):905–911, 1998.

[2] S. M. Abu-Sharkh, Z. F. Hussien, and J. K Sykulski. Current control of three-phase PWM inverters for embedded generators. In *Proceedings of the Eighth International Conference on Power Electronics and Variable Speed Drives*, pages 524–529, London, Sept. 2000.

[3] Hyo-Sung Ahn. *Robust and adaptive learning control design in the iteration domain*. PhD thesis, Utah State University, 2006.

[4] Hyun-Sik Ahn, Joong-Min Park, Do-Hyun Kim, Ick Choy, Joong-Ho Song, and Masayoshi Tomizuka. Extended direct learning control for multi-input multi-output nonlinear systems. In *Proceedings of IFAC 15th World Congress*, Barcelona, Spain, July 2002. IFAC.

[5] T. Al-Towaim, A. D. Barton, P. L. Lewin, E. Rogers, and D. H. Owens. Iterative learning control – 2D control systems from theory to application. *Int. J. of Control*, 77(9):877–893, 2004.

[6] T. Al Towaim, P. L. Lewin, and E. Rogers. Higher order ILC versus alternatives applied to chain conveyor systems. In *Proceedings of IFAC 15th World Congress*, Barcelona, Spain, July 2002. IFAC.

[7] P. Albertos, M. Olivares, and A. Sala. Fuzzy logic based look-up table controller with generalization. In *Proceedings of the 2000 American Control Conference*, pages 1949–1953, Chicago, IL, June 2000.

[8] G. Alefeld and J. Herzberger. *Introduction to Interval Computations*. Academic Press, New York, 1983.

[9] N. Amann, D. H. Owens, and E. Rogers. Predictive optimal iterative learning control. *Int. J. of Control*, 69(2):203–226, 1998.

[10] N. Amann, H. Owens, and E. Rogers. Iterative learning control for discrete-time systems with exponential rate of convergence. *IEE Proceedings – Control Theory and Applications*, 143(2):217–224, March 1996.

[11] Daniel Andres and Madhukar Pandit. Convergence and robustness of iterative learning control for strongly positive systems. In *Proceedings of*

the 3rd Asian Control Conference, pages 1866–1871, Shanghai, China, 2000. ASCC.

[12] Daniel Andres and Madhukar Pandit. Convergence and robustness of iterative learning control for strongly positive systems. *Asian Journal of Control*, 4(1):1–10, 2002.

[13] E. N. Antoniou and S. Vologiannidis. A new family of companion forms of polynomial matrices. *Electronic Journal of Linear Algebra*, 11:78–87, 2004.

[14] M. Arif and H. Inooka. Iterative manual control model of human operator. *Biological Cybernetics*, 81(5-6):445–455, 1999.

[15] M. Arif, T. Ishihara, and H. Inooka. Application of PILC to uncertain nonlinear systems for slowly varying desired trajectories. In *Proceedings of the 37th SICE Annual Conference*, pages 991–994, Chiba, Japan, July 1998.

[16] M. Arif, T. Ishihara, and H. Inooka. Iterative learning control using information database (ILCID). *J. of Intelligent & Robotic Systems*, 25(1):27–41, 1999.

[17] M. Arif, T. Ishihara, and H. Inooka. Iterative learning control utilizing the error prediction method. *J. of Intelligent & Robotic Systems*, 25(2):95–108, 1999.

[18] M. Arif, T. Ishihara, and H. Inooka. Using experience to get better convergence in iterative learning control. In *Proceedings of the 38th SICE Annual Conference*, pages 1211–1214, Morioka, Japan, July 1999.

[19] M. Arif, T. Ishihara, and H. Inooka. Prediction-based iterative learning control (PILC) for uncertain dynamic nonlinear systems using system identification technique. *J. of Intelligent & Robotic Systems*, 27(3):291–304, 2000.

[20] M. Arif, T. Ishihara, and H. Inooka. Incorporation of experience in iterative learning controllers using locally weighted learning. *Automatica*, 37(6):881–888, 2001.

[21] M. Arif, T. Ishihara, and H. Inooka. Experience-based iterative learning controllers for robotic systems. *Journal of Intelligent & Robotic Systems*, 35(4):381–396, 2002.

[22] M. Arif, T. Ishihara, and H. Inooka. A learning control for a class of linear time varying systems using double differential of error. *Journal of Intelligent & Roboctic Systems*, 36(2):223–234, 2003.

[23] Muhammad Arif, Tadashi Ishihara, and Hikaru Inooka. Model based iterative learning control (MILC) for uncertain dynamic non-linear systems. In *Proceedings of the 14th World Congress of IFAC*, pages 459–564, Beijing, China, 1999. IFAC.

[24] S. Arimoto. Equivalence of learnability to output-dissipativity and application for control of nonlinear mechanical systems. In *1999 IEEE International Conference on Systems, Man, and Cybernetics*, pages 39–44, Tokyo, Japan, Oct. 1999.

[25] S. Arimoto, H. Y. Han, P. T. A. Nguyen, and S. Kawamura. Iterative learning of impedance control from the viewpoint of passivity. *Int. J. of Robust and Nonlinear Control*, 10(8):597–609, 2000.

[26] S. Arimoto, S. Kawamura, and Hyun-Yong Han. System structure rendering iterative learning convergent. In *Proceedings of the 37th IEEE Conference on Decision and Control*, pages 672–677, Tampa, FL, Dec. 1998.

[27] S. Arimoto, S. Kawamura, and F. Miyazaki. Bettering operation of dynamic systems by learning: A new control theory for servomechanism or mechatronic systems. In *Proceedings of 23rd Conference on Decision and Control*, pages 1064–1069, Las Vegas, Nevada, December 1984.

[28] S. Arimoto, S. Kawamura, and F. Miyazaki. Bettering operation of robots by learning. *J. of Robotic Systems*, 1(2):123–140, 1984.

[29] S. Arimoto, S. Kawamura, and F. Miyazaki. Convergence, stability and robustness of learning control schemes for robot manipulators. In M. J. Jamishidi, L. Y. Luh, and M. Shahinpoor, editors, *Recent Trends in Robotics: Modelling, Control, and Education*, pages 307–316. Elsevier, New York, 1986.

[30] S. Arimoto and T. Naniwa. Equivalence relations between learnability, output-dissipativity and strict positive realness. *Int. J. of Control*, 73(10):824–831, 2000.

[31] S. Arimoto and T. Naniwa. Learnability and adaptability from the viewpoint of passivity analysis. *Intelligent Automation and Soft Computing*, 8(2):71–94, 2002.

[32] S. Arimoto, P. T. A. Nguyen, and T. Naniwa. Learning of robot tasks via impedance matching. In *Proceedings of the IEEE International Conference on Robotics and Automation*, pages 2786–2792, Detroit, MI May 1999.

[33] Suguru Arimoto and Pham Thuc Anh Nguyen. Iterative learning based on output-dissipativity. In *Proceedings of the 3rd Asian Control Conference*, Shanghai, China, 2000. ASCC.

[34] K. E. Avrachenkov, H. S. M. Beigi, and R. W. Longman. Updating procedures for iterative learning control in Hilbert space. *Intelligent Automation and Soft Computing*, 8(2):183–189, 2002.

[35] K.E. Avrachenkov, H.S.M. Beigi, and R.W. Longman. Updating procedures for iterative learning control in Hilbert space. In *Proceedings of the 38th IEEE Conference on Decision and Control*, pages 276–280, Phoenix, AZ Dec. 1999.

[36] A Azenha. Iterative learning in variable structure position/force hybrid control of manipulators. *Robotica*, 18:213–217 Part 2, 2000.

[37] B. Ross Barmish. *New Tools for Robustness of Linear Systems*. Macmillan Publishing Company, New York, 1994.

[38] A. D. Barton, P. L. Lewin, and D. J. Brown. Practical implementation of a real-time iterative learning position controller. *Int. J. of Control*, 73(10):992–999, 2000.

[39] P. Batra. On necessary conditions for real robust Schur-stability. *IEEE Trans. on Automatic Control*, 48(2):259–261, 2003.

[40] Homayoon S. M. Beigi (Ed.). Special Issue: learning and repetitive control. *Intelligent Automation and Soft Computing*, 8(2), 2002.

[41] L. Ben-brahim, M. Benammar, and M. A. Alhamadi. A new iterative learning control method for PWM inverter current regulation. In *Proceedings of the Fifth International Conference on Power Electronics and Drive Systems*, pages 1460–1465, Nov. 2003.

[42] S. P. Bhattacharyya, H. Chapellat, and L. H. Keel. *Robust Control: The Parameter Approach*. Prentice Hall, 1995.

[43] Z. Bien and K. M. Huh. Higher-order iterative learning control algorithm. In *IEE Proceedings Part D, Control Theory and Applications*, pages 105–112, May 1989.

[44] Zeungnam Bien and Jian-Xin Xu. *Iterative Learning Control – Analysis, Design, Integration and Applications*. Kluwer Academic Publishers, 1998.

[45] E. Bonanomi, J. Sefcik, M. Morari, and M. Morbidelli. Analysis and control of a turbulent coagulator. *Industrial & Engineering Chemistry Research*, 43(19):6112–6124, 2004.

[16] Michael S. Branicky, Stephen M. Philips, and Wei Zhang. Stability of networked control systems: explicit analysis of delay. In *Proc. of the American Control Conference*, pages 2352–2357, Chicago, IL, 2000. ACC.

[47] Douglas A. Bristow, Marina Tharayil, and Andrew G. Alleyne. A survey of iterative learning control: A learning-based method for high-performance tracking control. *IEEE Control Systems Magazine*, 26(3):96–114, 2006.

[48] C. C. Cheah. Robustness of time-scale learning of robot motions to uncertainty in acquired knowledge. *J. of Robotic Systems*, 18(10):599–608, 2001.

[49] C. C. Cheah and D. W. Wang. Learning impedance control for robotic manipulators. *IEEE Trans. on Robotics and Automation*, 14(3):452–465, 1998.

[50] C. K. Chen and W. C. Zeng. The iterative learning control for the position tracking of the hydraulic cylinder. *JSME Int. J. Series C-Mechanical Systems Machine Elements and Manufacturing*, 46(2):720–726, 2003.

[51] Chi-Tsong Chen. *Linear System Theory and Design*. Saunders College Publishing, Philadelphia, PA, 1984.

[52] Chi-Tsong Chen. *Linear System Theory and Design*. Oxford University Press, 1999.

[53] Chih-Keng Chen and James Hwang. PD-type iterative learning control for trajectory tracking of a pneumatic X-Y table with disturbances. In *Proceedings of the 2004 IEEE International Conference on Robotics and Automation*, pages 3500–3505, April 26–May 1 2004.

[54] H. F. Chen. Almost sure convergence of iterative learning control for stochastic systems. *Science in China Series F-Information Sciences*, 46(1):67–79, 2003.

[55] H. F. Chen and H. T. Fang. Output tracking for nonlinear stochastic systems by iterative learning control. *IEEE Trans. on Automatic Control*, 49(4):583–588, 2004.

[56] Huadong Chen and Ping Jiang. Adaptive iterative feedback control for nonlinear system with unknown high-frequency gain. In *Proceedings of the 4th World Congress on Intelligent Control and Automation*, pages 847–851, June 2002.

[57] W. Chen, Y.C. Soh, and C.W. Yin. Iterative learning control for constrained nonlinear systems. *Int. J. of Systems Science*, 30(6):659–664, 1999.

[58] Y. Chen, C. Wen, J.-X. Xu, and M. Sun. An initial state learning method for iterative learning control of uncertain time-varying systems. In *Proc. of the 35th IEEE CDC*, pages 3996–4001, Kobe, Japan, Dec. 1996.

[59] Y. Q. Chen and K. L. Moore. Comments on United States Patent 3,555,252–Learning Control of Actuators in Control Systems. In *Proceedings of the 2000 International Conference on Automation, Robotics, and Control*, Singapore, Dec. 2000.

[60] Y. Q. Chen and K. L. Moore. On D^α type iterative learning control. In *Proceedings of the 40th IEEE Conference on Decision and Control*, pages 4451–4456, Orlando, FL, Dec. 2001.

[61] Y. Q. Chen and K. L. Moore. Harnessing the nonrepetitiveness in iterative learning control. In *Proceedings of the 41st IEEE Conference on Decision and Control*, pages 3350–3355, Las Vegas, Nevada, Dec. 2002.

[62] Y. Q. Chen and K. L. Moore. An optimal design of PD-type iterative learning control with monotonic convergence. In *Proceedings of the 2002 IEEE International Symposium on Intelligent Control*, pages 55–60, Vancouver, Canada, Oct. 2002.

[63] Y. Q. Chen and K. L. Moore. PI-type iterative learning control revisited. In *Proceedings of the 2002 American Control Conference*, pages 2138–2143, May 2002.

[64] Y. Q. Chen and K. L. Moore. Iterative learning control with iteration-domain adaptive feedforward compensation. In *Proceedings of the 42nd IEEE Conference on Decision and Control*, pages 4416–4421, Maui, HI, Dec. 2003.

[65] Y. Q. Chen, K. L. Moore, and V. Bahl. Learning feedforward control using a dilated B-spline network: Frequency domain analysis and design. *IEEE Trans. on Neural Networks*, 15(2):355–366, 2004.

[66] Y. Q. Chen and C. Wen. *Iterative Learning Control: Convergence, Robustness and Applications*, volume LNCIS-248 of *Lecture Notes series on Control and Information Science*. Springer-Verlag, London, 1999.

[67] Y. Q. Chen, C. Y. Wen, J. X. Xu, and M. X. Sun. High-order iterative learning identification of projectile's aerodynamic drag coefficient curve

from radar measured velocity data. *IEEE Trans. on Control Systems Technology*, 6(4):563–570, 1998.

[68] YangQuan Chen. *High-order Iterative Learning Control: Convergence, Robustness and Applications*. PhD thesis, Nanyang Technological University, Singapore, 1997.

[69] YangQuan Chen, Huifang Dou, and Kok-Kiong Tan. Iterative learning control via weighted local-symmetrical-integration. *Asian Journal of Control*, 3(4):352–356, 2001.

[70] YangQuan Chen and Kevin L. Moore. Improved path following for an omni-directional vehicle via practical iterative learning control using local symmetrical double-integration. In *Proceedings of the 3rd Asian Control Conference*, Shanghai, China, 2000. ASCC.

[71] YangQuan Chen and Kevin L. Moore. A practical iterative learning path-following control of an omni-directional vehicle. *Asian Journal of Control*, 4(1):90–98, 2002.

[72] YangQuan Chen, LeeLing Tan, KianKeong Ooi, Qiang Bi, and KokHiang Cheong. Repeatable runout compensation using a learning algorithm with scheduled parameters. United States Patent 6,437,936, August 2002.

[73] Yangquan Chen, Jian-Xin Xu, and Mingxuan Sun. Extracting aerobomb's aerodynamic drag coefficient curve from theodolite data via iterative learning. In *Proceedings of the 14th World Congress of IFAC*, pages 115–120, Beijing, China, 1999. IFAC.

[74] Y.Q. Chen, Z.M. Gong, and C.Y. Wen. Analysis of a high-order iterative learning control algorithm for uncertain nonlinear systems with state delays. *Automatica*, 34(3):345–353, 1998.

[75] Y.Q. Chen, C. Wen, Z. Gong, and M. Sun. An iterative learning controller with initial state learning. *IEEE Trans. on Automatic Control*, 44(2):371–376, 1999.

[76] H. G. Chiacchiarini and P. S. Mandolesi. Unbalance compensation for active magnetic bearings using ILC. In *Proceedings of the 2001 IEEE International Conference on Control Applications*, pages 58–63, Mexico City, Mexico, Sept. 2001.

[77] C. J. Chien. A discrete iterative learning control for a class of nonlinear time-varying systems. *IEEE Trans. on Automatic Control*, 43(5):748–752, 1998.

[78] C. J. Chien, C. T. Hsu, and C. Y. Yao. Fuzzy system-based adaptive iterative learning control for nonlinear plants with initial state errors. *IEEE Trans. on Fuzzy Systems*, 12(5):724–732, 2004.

[79] C. J. Chien and C. Y. Yao. Iterative learning of model reference adaptive controller for uncertain nonlinear systems with only output measurement. *Automatica*, 40(5):855–864, 2004.

[80] C. J. Chien and C. Y. Yao. An output-based adaptive iterative learning controller for high relative degree uncertain linear systems. *Automatica*, 40(1):145–153, 2004.

[81] Chiang-Ju Chien. An output based iterative learning controller for systems with arbitrary relative degree. In *Proceedings of the 3rd Asian Control Conference*, Shanghai, China, 2000. ASCC.

[82] Chiang-Ju Chien. A sampled-data iterative learning control using fuzzy network design. *Int. J. of Control*, 73(10):902–913, 2000.

[83] Chiang-Ju Chien and Li-Chen Fu. A neural network based learning controller for robot manipulators. In *Proceedings of the 39th IEEE Conference on Decision and Control*, pages 1748–1753, Sydney, NSW, Dec. 2000.

[84] Chiang-Ju Chien and Li-Chen Fu. An iterative learning control of nonlinear systems using neural network design. *Asian Journal of Control*, 4(1):21–29, 2002.

[85] C. J. Chien. A sampled-data iterative learning control using fuzzy network design. *Int. J. of Control*, 73(10):902–913, 2000.

[86] I. Chin, S. J. Qin, K. S. Lee, and M. Cho. A two-stage iterative learning control technique combined with real-time feedback for independent disturbance rejection. *Automatica*, 40(11):1913–1920, 2004.

[87] In Sik Chin, Jinho Lee, Hyojin Ahn, Sangrae Joo, K. S. Lee, and Daeryook Yang. Optimal iterative learning control of wafer temperature uniformity in rapid thermal processing. In *Proceedings of the IEEE International Symposium on Industrial Electronics*, pages 1225–1230, Pusan, South Korea, June 2001.

[88] S. Chin, Kwang S. Lee, and Jay H. Lee. A unified framework for combined quality and tracking control of batch processes. In *Proceedings of the 14th World Congress of IFAC*, pages 349–354, Beijing, China, 1999. IFAC.

[89] M. Cho, Y. Lee, S. Joo, and K. S. Lee. Sensor location, identification, and multivariable iterative learning control of an RTP process for maximum uniformity of wafer temperature distribution. In *Proceedings of the 11th IEEE International Conference on Advanced Thermal Processing of Semiconductors*, pages 177–184, Sept. 2003.

[90] C. H. Choi and G. M. Jeong. Perfect tracking for maximum-phase nonlinear systems by iterative learning control. *Int. J. of Systems Science*, 32(9):1177–1183, 2001.

[91] Chong-Ho Choi and Tae-Jeong Jang. Iterative learning control in feedback systems based on an objective function. *Asian Journal of Control*, 2(2):101–110, 2000.

[92] Chong-Ho Choi and Gu-Min Jeong. Iterative learning control for linear nonminimum-phase systems based on initial state learning. In *Proceedings of the 3rd Asian Control Conference*, Shanghai, China, 2000. ASCC.

[93] Chong-Ho Choi, Gu-Min Jeong, and Tae-Jeong Jang. Optimal output tracking for discrete time nonlinear systems. In *Proceedings of the 14th World Congress of IFAC*, pages 1–6, Beijing, China, 1999. IFAC.

[94] J. Y. Choi and H. M. Do. A learning approach of wafer temperature control in a rapid thermal processing system. *IEEE Trans. on Semiconductor Manufacturing*, 14(1):1–10, 2001.

[95] Joon-Young Choi, Jiho Uh, and Jin S. Lee. Iterative learning control of robot manipulator with I-type parameter estimator. In *Proceedings of the 2001 American Control Conference*, pages 646–651, Arlington, VA, June 2001.

[96] J. Y. Choi and J. S. Lee. Adaptive iterative learning control of uncertain robotic systems. *IEE Proceedings - Control Theory and Applications*, 147(2):217–223, 2000.

[97] J. Y. Choi and H. J. Park. Use of neural networks in iterative learning control systems. *Int. J. of Systems Science*, 31(10):1227–1239, 2000.

[98] Jyh-Horng Chou. Pole-assignment robustmess in a specified disk. *System and Control Letters*, 16:41–44, 1991.

[99] T. W. S. Chow and Y. Fang. An iterative learning control method for continuous-time systems based on 2–D system theory. *IEEE Trans. on Circuits and Systems I – Fundamental Theory and Applications*, 45(6):683–689, 1998.

[100] T. W. S. Chow, X. D. Li, and Y. Fang. A real-time learning control approach for nonlinear continuous-time system using recurrent neural networks. *IEEE Trans. on Industrial Electronics*, 47(2):478–486, 2000.

[101] B. T. Costic, M. S. de Queiroz, and D. N. Dawson. A new learning control approach to the active magnetic bearing benchmark system. In *Proceedings of the 2000 American Control Conference*, pages 2639–2643, Chicago, IL, June 2000.

[102] D. de Roover. Synthesis of a robust iterative learning controller using an H_∞ approach. In *Proceedings of the 35th IEEE Conference on Decision and Control*, Kobe, Japan, December 1996.

[103] D. de Roover and Ollo H. Bosgra. Synthesis of robust multivariable iteraive learning controller with application to a wafer stage motion system. *International Journal of Control*, 73(10):968–979, 2000.

[104] Carlos E. de Souza, Minyue Fu, and Lihua Xie. H_∞ analysis and synthesis of discrete-time systems with time varying uncertainty. *IEEE Transactions on Automatic Control*, 38(3):459–462, 1993.

[105] D. Lj. Debeljković, M. Aleksendrić, N. Yi-Yong, and Q. L. Zhang. Lyapunov and non-Lyapunov stability of linear discrete time delay systems. *FACTA Universitatis Series: Mechanical Engineering*, 1(9):1147–1160, 2002.

[106] A. S. Deif. Singular values of an interval matrix. *Linear Algebra and Its Applications*, 151:125–134, 1991.

[107] David F. Delchamps. Analytic feedback control and the algebraic Riccati equation. *IEEE Trans. on Automatic Control*, 29(11):1031–1033, 1984.

[108] J. J. D. Delgado-Romero, R. S. Gonzalez-Garza, J. A. Rojas-Estrada, G. Acosta-Villarreal, and F. Delgado-Romero. Some very simple Hurwitz

and Schur stability tests for interval matrices. In *Proceedings of the 30th Conference on Decision and Control*, pages 2980–2981, Kobe, Japan, December 1996. IEEE.

[109] Liu Deman, A. Konno, and M. Uchiyama. Flexible manipulator trajectory learning control with input preshaping method. In *38th Annual Conference Proceedings of the SICE*, pages 967–972, Morioka, Japan, July 1999.

[110] B. G. Dijkstra and O. H. Bosgra. Convergence design considerations of low order Q-ILC for closed loop systems, implemented on a high precision wafer stage. In *Proceedings of the 41st IEEE Conference on Decision and Control*, pages 2494–2499, Las Vegas, Nevada, Dec. 2002.

[111] B. G. Dijkstra and O. H. Bosgra. Extrapolation of optimal lifted system ILC solution, with application to a waferstage. In *Proceedings of the 2002 American Control Conference*, pages 2595–2600, Anchorage, AK, May 2002.

[112] B. G. Dijkstra and O. H. Bosgra. Noise suppression in buffer-state iterative learning control, applied to a high precision wafer stage. In *Proceedings of the 2002 International Conference on Control Applications*, pages 998–1003, Glasgow, U.K, Sept. 2002.

[113] B. G. Dijkstra and O. H. Bosgra. Exploiting iterative learning control for input shaping, with application to a wafer stage. In *Proceedings of the 2003 American Control Conference*, pages 4811–4815, Denver, CO, June 2003.

[114] Branko Dijkstra. *Iterative learning control with applications to a waferstage*. PhD thesis, Delft University of Technology (Netherlands), 2004.

[115] T.Y. Doh and J.H. Moon. An iterative learning control for uncertain systems using structured singular value. *Journal of Dynamic Systems Measurement and Control – Trans. of The ASME*, 121(4):660–667, 1999.

[116] T.Y. Doh, J.H. Moon, K.B. Jin, and M.J. Chung. Robust iterative learning control with current feedback for uncertain linear systems. *Int. J. of Systems Science*, 30(1):39–47, 1999.

[117] H. F. Dou, K. K. Tan, T. H. Lee, and Z. Y. Zhou. Iterative learning feedback control of human limbs via functional electrical stimulation. *Control Engineering Practice*, 7(3):315–325, 1999.

[118] B. J. Driessen and N. Sadegh. Convergence theory for multi-input discrete-time iterative learning control with Coulomb friction, continuous outputs, and input bounds. In *Proceedings of IEEE SoutheastCon*, pages 287–293, Columbia, SC, April 2002.

[119] B. J. Driessen and N. Sadegh. Global convergence for two-pulse rest-to-rest learning for single-degree-of-freedom systems with stick-slip Coulomb friction. In *Proceedings of the 41st IEEE Conference on Decision and Control*, pages 3338–3343, Las Vegas, Nevada, Dec. 2002.

[120] B. J. Driessen and N. Sadegh. Multi-input square iterative learning control with bounded inputs. *Journal of Dynamic Systems Measurement and Control – Trans. of the ASME*, 124(4):582–584, 2002.

[121] B. J. Driessen and N. Sadegh. Convergence theory for multi-input discrete-time iterative learning control with Coulomb friction, continuous outputs, and input bounds. *Int. J. of Adaptive Control and Signal Processing*, 18(5):457–471, 2004.

[122] M. Dymkov, I. Gaishun, K. Galkowski, E. Rogers, and D. H. Owens. A Volterra operator approach to the stability analysis of a class of 2D linear systems. In *Proceedings of the 2001 European Control Conference*, Seminário de Vilar, Porto, Portugal, Sept. 2001. ECC.

[123] M. Dymkov, I. Gaishun, E. Rogers, K. Galkowski, and D. H. Owens. On the observability properties of a class of 2D discrete linear systems. In *Proceedings of the 40th IEEE Conference on Decision and Control*, pages 3625–3630, Orlando, FL, Dec. 2000.

[124] M. Dymkov, I. Gaishun, E. Rogers, K. Galkowski, and D. H. Owens. z-transform and volterra-operator based approaches to controllability and observability analysis for discrete linear repetitive processes. *Multidimensional Systems and Signal Processing*, 14(4):365–395, 2003.

[125] J. B. Edwards. Stability problems in the control of linear multipass processes. *Proc. IEE*, 121(11):1425–1431, 1974.

[126] J. B. Edwards and D. H. Owens. *Analysis and Control of Multipass Processes*. Research Studies Press, Taunton, Chichestor, 1982.

[127] H. Elci, R. W. Longman, M. Q. Phan, J. N. Juang, and R. Ugoletti. Simple learning control made practical by zero-phase filtering: Applications to robotics. *IEEE Trans. on Circuits and Systems I – Fundamental Theory and Applications*, 49(6):753–767, 2002.

[128] Y. Fang and T. W. S. Chow. Iterative learning control of linear discrete-time multivariable systems. *Automatica*, 34(11):1459–1462, 1998.

[129] Y. Fang and T. W. S. Chow. 2–D analysis for iterative learning controller for discrete-time systems with variable initial conditions. *IEEE Trans. on Circuits and Systems I – Fundamental Theory and Applications*, 50(5):722–727, 2003.

[130] Y. Fang and T.W.S. Chow. Counterexample to iterative learning control of linear discrete-time multivariable systems – Author's reply. *Automatica*, 36(2): 329–329, 2000.

[131] Yong Fang, Yeng Chai Soh, and G. G. Feng. Convergence analysis of iterative learning control with uncertain initial conditions. In *Proceedings of the 4th World Congress on Intelligent Control and Automation*, pages 960–963, Shanghai, China, June 2002.

[132] M. French, G. Munde, E. Rogers, and D.H. Owens. Recent developments in adaptive iterative learning control. In *Proceedings of the 38th IEEE Conference on Decision and Control*, pages 264–269, Phoenix, AZ, Dec. 1999.

[133] M. French and E. Rogers. Nonlinear iterative learning by an adaptive Lyapunov technique. In *Proceedings of the 37th IEEE Conference on Decision and Control*, pages 175–180, Tampa, FL, Dec. 1998.

[134] M. French and E. Rogers. Non-linear iterative learning by an adaptive Lyapunov technique. *Int. J. of Control*, 73(10):840–850, 2000.

[135] M. French, E. Rogers, H. Wibowo, and D. H. Owens. A 2D systems approach to iterative learning control based on nonlinear adaptive control techniques. In *Proceedings of the 2001 IEEE International Symposium on Circuits and Systems*, pages 429–432, Sydney, NSW, May 2001.

[136] James A. Frueh and Minh Q. Phan. Linear quadratic optimal learning control (LQL). *Int. J. of Control*, 73(10):832–839, 2000.

[137] James Arthur Frueh. *Iterative learning control with basis functions*. PhD thesis, Princeton University, 2000.

[138] K. Fujimoto, H. Kakiuchi, and T. Sugie. Iterative learning control of Hamiltonian systems. In *Proceedings of the 41st IEEE Conference on Decision and Control*, pages 3344–3349, Las Vegas, NV, Dec. 2002.

[139] K. Fujimoto and T. Sugie. Iterative learning control of Hamiltonian systems based on self-adjoint structure-I/O based optimal control. In *Proceedings of the 41st SICE Annual Conference*, pages 2573–2578, Aug. 2002.

[140] K. Fujimoto and T. Sugie. Iterative learning control of Hamiltonian systems: I/O based optimal control approach. *IEEE Trans. on Automatic Control*, 48(10):1756–1761, 2003.

[141] Kenji Fujimoto and Toshiharu Sugie. On adjoints of Hamiltonian systems. In *Proceedings of IFAC 15th World Congress*, Barcelona, Spain, July 2002. IFAC.

[142] K. Furuta and S. Phoojaruenchanachai. An algebraic approach to discrete-time H_∞ control problems. In *Proc. American Control Conference*, pages 3067–3072, San Diego, CA, May 1990. AACC.

[143] K. Furuta and M. Yamakita. The design of a learning control system for multivariable systems. In *Proceedings of the 30th Conference on Decision and Control*, pages 371–376, Philadelphia, PA, 1987.

[144] K. Galkowski, J. Lam, E. Rogers, S. Xu, B. Sulikowski, W. Paszke, and D. H. Owens. LMI based stability analysis and robust controller design for discrete linear repetitive processes. *Int. J. of Robust and Nonlinear Control*, 13(13):1195–1211, 2003.

[145] K. Galkowski, W. Paszke, E. Rogers, and D. H. Owens. Stabilization and robust control of metal rolling modeled as a 2D linear system. In *Proceedings of IFAC 15th World Congress*, Barcelona, Spain, July 2002. IFAC.

[146] K. Galkowski, W. Paszke, B. Sulikowski, E. Rogers, and D. H. Owens. LMI based stability analysis and controller design for a class of 2D continuous-discrete linear systems. In *Proceedings of the 2002 American Control Conference*, pages 29–34, Anchorage, AK, May 2002.

[147] K. Galkowski, E. Rogers, A. Gramacki, J. Gramacki, and D. H. Owens. Stability and dynamic boundary condition decoupling analysis for a class of 2-D discrete linear systems. *IEE Proceedings – Circuits Devices and Systems*, 148(3):126–134, 2001.

[148] K. Galkowski, E. Rogers, and D.H. Owens. New 2D models and a transition matrix for discrete linear repetitive processes. *Int. J. of Control,* 72(15):1365–1380, 1999.

[149] K. Galkowski, E. Rogers, J. Wood, S. E. Benton, and D. H. Owens. One-dimensional equivalent model and related approaches to the analysis of discrete nonunit memory linear repetitive processes. *Circuits Systems and Signal Processing,* 21(6):525–534, 2002.

[150] K. Galkowski, E. Rogers, S. Xu, J. Lam, and D. H. Owens. LMIs – a fundamental tool in analysis and controller design for discrete linear repetitive processes. *IEEE Trans. on Circuits and Systems I – Fundamental Theory and Applications,* 49(6):768–778, 2002.

[151] F. R. Gao, Y. Yang, and C. Shao. Robust iterative learning control with applications to injection molding process. *Chemical Engineering Science,* 56(24):7025–7034, 2002.

[152] S. S. Garimella and K. Srinivasan. Application of iterative learning control to coil-to-coil control in rolling. *IEEE Trans. on Control Systems Technology,* 6(2):281–293, 1998.

[153] Arthur Gelb. *Applied Optimal Estimation.* The MIT Press, 1974.

[154] J. Ghosh and B. Paden. Iterative learning control for nonlinear nonminimum phase plants with input disturbances. In *Proceedings of the 1999 American Control Conference,* pages 2584–2589, San Diego, CA, June 1999.

[155] J. Ghosh and B. Paden. Pseudo-inverse based iterative learning control for nonlinear plants with disturbances. In *Proceedings of the 38th IEEE Conference on Decision and Control,* pages 5206–5212, Phoenix, AZ, Dec 1999.

[156] J. Ghosh and B. Paden. Pseudo-inverse based iterative learning control for plants with unmodelled dynamics. In *Proceedings of the 2000 American Control Conference,* pages 472–476, Chicago, IL, June 2000.

[157] J. Ghosh and B. Paden. Iterative learning control for nonlinear nonminimum phase plants. *Journal of Dynamic Systems Measurement and Control – Trans. of the ASME,* 123(1):21–30, 2001.

[158] J. Ghosh and B. Paden. A pseudo-inverse-based iterative learning control. *IEEE Trans. on Automatic Control,* 45(5):831–837, 2002.

[159] Jayati Ghosh. *Tracking of periodic trajectories in nonlinear plants.* PhD thesis, University of California, Santa Barbara, 2000.

[160] I. Gohberg, P. Lancaster, and L. Rodman. *Matrix Polynomials.* Academic Press, New York, 1982.

[161] P. B. Goldsmith. The equivalence of LTI iterative learning control and feedback control. In *Proceedings of the 2000 IEEE International Conference on Systems, Man, and Cybernetics,* pages 3443–3448, Nashville, TN, Oct. 2001.

[162] P. B. Goldsmith. On the equivalence of causal LTI iterative learning control and feedback control. *Automatica,* 38(4):703–708, 2002.

[163] P. B. Goldsmith. Author's reply to Comments on 'On the equivalence of causal LTI iterative learning control and feedback control'. *Automatica*, 40(5):899–900, 2004.

[164] P.B. Goldsmith. The fallacy of causal iterative learning control. In *Proceedings of the 40th IEEE Conference on Decision and Control*, pages 4475–4480, Orlando, FL, Dec. 2001.

[165] Peter B. Goldsmith. Stability, convergence, and feedback equivalence of LTI iterative learning control. In *Proceedings of IFAC 15th World Congress*, Barcelona, Spain, July 2002. IFAC.

[166] S. Gopinath and I. N. Kar. Iterative learning control scheme for manipulators including actuator dynamics. *Mechanism and Machine Theory*, 39(12):1367–1384, 2004.

[167] D. Gorinevsky. Distributed system loopshaping design of iterative control for batch processes. In *Proceedings of the 38th IEEE Conference on Decision and Control*, pages 245–250, Phoenix, AZ, Dec. 1999.

[168] Dimitry M. Gorinevsky. Iterative learning update for batch mode processing. United States Patent 6,647,354, November 2003.

[169] J. Gramacki, A. Gramacki, K. Galkowski, E. Rogers, and D.H. Owens. MATLAB based tools for 2D linear systems with application to iterative learning control schemes. In *Proceedings of the 1999 IEEE International Symposium on Computer Aided Control System Design*, pages 410–415, Kohala Coast, HI, Aug. 1999.

[170] M. Grundelius and B. Bernhardsson. Constrained iterative learning control of liquid slosh in an industrial packaging machine. In *Proceedings of the 39th IEEE Conference on Decision and Control*, pages 4544–4549, Sydney, Australia, Dec. 2000.

[171] S. Gunnarsson and M. Norrlöf. Some aspects of an optimization approach to iterative learning control. In *Proceedings of the 38th IEEE Conference on Decision and Control*, pages 1581–1586, Phoenix, AZ, Dec 1999.

[172] Sy-Ming Guu and Chin Tzong Pang. On the convergence to zero of infinite products of interval matrices. *J. Matrix Anal. Appl*, 25(3):739–751, 2004.

[173] K. Hamamoto and T. Sugie. An iterative learning control algorithm within prescribed input-output subspace. *Automatica*, 37(11): 1803–1809, 2001.

[174] K. Hamamoto and T. Sugie. Iterative learning control for robot manipulators using the finite dimensional input subspace. In *Proceedings of the 40th IEEE Conference on Decision and Control*, pages 4926–4931, Orlando, FL, Dec. 2001.

[175] K. Hamamoto and T. Sugie. Iterative learning control for robot manipulators using the finite dimensional input subspace. *IEEE Trans. on Robotics and Automation*, 18(4):632–635, 2002.

[176] Kenichi Hamamoto and Toshiharu Sugie. An iterative learning control algorithm within prescribed input-output subspace. In *Proceedings of*

the 14th World Congress of IFAC, pages 501–506, Beijing, China, 1999. IFAC.

[177] H. S. Han and J. G. Lee. Necessary and sufficient conditions for stability of time varying discrete interval matrices. *Int. J. Control*, 59:1021–1029, 1994.

[178] Elden R. Hansen and G. William Walster. *Global Optimization Using Interval Analysis*. Marcel Dekker, New York, 2003.

[179] G. I. Hargreaves. Interval analysis in MATLAB. Numerical Analysis Report No. 416, The University of Manchester, December 2002.

[180] J. Hätönen, K. L. Moore, and D. H. Owens. An algebraic approach to iterative learning control. In *Proceedings of the 2002 IEEE International Symposium on Intelligent Control*, pages 37–42, Oct. 2002.

[181] J. Hätönen and D. H. Owens. Convex modifications to an iterative learning control law. *Automatica*, 40(7):1213–1220, 2004.

[182] J. J. Hätönen, K. Feng, and D. H. Owens. New connections between positivity and parameter-optimal iterative learning control. In *Proceedings of the 2003 IEEE International Symposium on Intelligent Control*, pages 69–74, Houston, Texas, Oct. 2003.

[183] J. J. Hätönen, T. J. Harte, D. H. Owens, J. D. Ratcliffe, P. L. Lewin, and E. Rogers. A new robust iterative learning control algorithm for application on a gantry robot. In *Proceedings of the IEEE Conference Emerging Technologies and Factory Automation*, pages 305 312, Sept. 2003.

[184] Jari Hätönen. *Issues of algebra and optimality in iterative learning control*. PhD thesis, University of Oulu (Finland), 2004.

[185] V. Hatzikos, J. Hätönen, and D. H. Owens. Genetic algorithms in norm-optimal linear and non-linear iterative learning control. *Int. J. of Control*, 77(2):188–197, 2004.

[186] V. Hatzikos and D. H. Owens. A genetic algorithm based optimisation method for iterative learning control systems. In *Proceedings of the Third International Workshop on Robot Motion and Control*, pages 423–428, Nov. 2002.

[187] V. E. Hatzikos, J. Hätönen, T. Harte, and D. H. Owens. Robust analysis of a genetic algorithm based optimization method for real-time iterative learning control applications. In *Proceedings of the IEEE Conference on Emerging Technologies and Factory Automation, 2003*, pages 396–401, Sept. 2003.

[188] V. E. Hatzikos, D. H. Owens, and J. Hätönen. An evolutionary based optimisation method for nonlinear iterative learning control systems. In *Proceedings of the 2003 American Control Conference*, pages 3638–3643, Denver, CO, June 2003.

[189] H. Havlicsek and A. Alleyne. Nonlinear control of an electrohydraulic injection molding machine via iterative adaptive learning. *IEEE/ASME Trans. on Mechatronics*, 4(3):312–323, 1999.

[190] H. Hengen, S. Hillenbrand, and M. Pandit. Algorithms for iterative learning control of nonlinear plants employing time-variant system descriptions. In *Proceedings of the 2000 IEEE International Conference on Control Applications*, pages 570–575, Anchorage, AK, Sept. 2000.

[191] D. Henrion, D. Arzelier, D. Peaucelle, and M. Sebek. An LMI condition for robust stability of polynomial matrix polytopes. *Automatica*, 37(3):461–468, 2001.

[192] David Hertz. The extreme eigenvalues and stability of real symmetric interval matrices. *IEEE Trans. on Automatic Control*, 37(4):532–535, 1992.

[193] L. Hideg. Design of a static neural element in an iterative learning control scheme. In *Proceedings of the 37th IEEE Conference on Decision and Control*, pages 690–694, Tampa, FL, Dec. 1998.

[194] Nicholas J. Higham and Françoise Tisseur. Bounds for eigenvalues of matrix polynomials. *Linear Algebra and Its Applications*, 358:5–22, 2003.

[195] S. Hillenbrand and M. Pandit. A discrete-time iterative learning control law with exponential rate of convergence. In *Proceedings of the 38th IEEE Conference on Decision and Control*, pages 1575–1580, Phoenix, AZ, Dec. 1999.

[196] Stefan Hillenbrand and Madhukar Pandit. An iterative learning controller with reduced sampling rate for plants with variations of initial states. *Int. J. of Control*, 73(10):882–889, 2000.

[197] K. J. G. Hinnen, R. Fraanje, and M. Verhaegen. The application of initial state correction in iterative learning control and the experimental validation on a piezoelectric tube scanner. *Proceedings of the Institution of Mechanical Engineers PT I – Journal of Systems and Control Engineering*, 218(16):503–511, 2004.

[198] W. Hoffmann, K. Peterson, and A. G. Stefanopoulou. Iterative learning control for soft landing of electromechanical valve actuator in camless engines. *IEEE Trans. on Control Systems Technology*, 11(2):174–184, 2003.

[199] W. Hoffmann and A. G. Stefanopoulou. Iterative learning control of electromechanical camless valve actuator. In *Proceedings of the 2001 American Control Conference*, pages 2860–2866, Arlington, VA, June 2001.

[200] Roger A. Horn and Charles R. Johnson. *Matrix Analysis*. Cambridge University Press, New York, 1985.

[201] Zhongsheng Hou and Jian-Xin Xu. Freeway traffic density control using iterative learning control approach. In *Proceedings of the 2003 IEEE Conference on Intelligent Transportation Systems*, pages 1081–1086, Oct. 2003.

[202] Mau Hsiang Shih, Yung Yih Lur, and Chin Tzong Pang. An ineqality for the spectral radius of an interval matrix. *Linear Algebra and Its Applications*, 274:27–36, 1998.

[203] Chun-Te Hsu, Chiang-Ju Chien, and Chia-Yu Yao. A new algorithm of adaptive iterative learning control for uncertain robotic systems. In *Proceedings of the IEEE International Conference on Robotics and Automation*, pages 4130–4135, Sept. 2003.

[204] Ai-Ping Hu and Nader Sadegh. Non-collocated control of a flexible-link manipulator tip using a multirate repetitive learning controller. In *Proceedings of the 3rd Mechatronics Forum International Conference*, Atlanta, Georgia, Sept. 2000.

[205] M. Hu, H. J. Du, S. F. Ling, Z. Y. Zhou, and Y. Li. Motion control of an electrostrictive actuator. *Mechatronics*, 14(2):153–161, 2004.

[206] M. Hu, Z. Y. Zhou, Y Li, and H. J. Du. Development of a linear electrostrictive servo motor. *Precision Engineering – Journal of the International Societies for Precision Engineering and Nanotechnolgy*, 25(4):316–320, 2001.

[207] Q. P. Hu, J. X. Xu, and T. H. Lee. Iterative learning control design for Smith predictor. *Systems & Control Letters*, 44(3):201–210, 2001.

[208] S. N. Huang and S. Y. Lim. Predictive iterative learning control. *Intelligent Automation and Soft Computing*, 9(2):103–112, 2003.

[209] S. N. Huang, K. K. Tan, and T. H. Lee. Iterative learning algorithm for linear time-varying with a quadratic criterion systems. *Proceedings of the Institution of Mechanical Engineers Part I Journal of Systems and Control Engineering*, 216(I3):309–316, 2002.

[210] Weiqing Huang. *Tracking control of nonlinear mechanical systems using Fourier series based learning control*. PhD thesis, Hong Kong Univ. of Sci. and Tech. (China), 1999.

[211] Y. C. Huang, M. Chan, Y. P. Hsin, and C. C. Ko. Use of PID and iterative learning controls on improving intra-oral hydraulic loading system of dental implants. *JSME Int. Journal Series C – Mechanical Systems Machine Elements and Manufacturing*, 46(4):1449–1455, 2003.

[212] Yi-Cheng Huang, Manuel Chan, Yi-Ping Hsin, and Ching-Chang Ko. Use of iterative learning control on improving intra-oral hydraulic loading system of dental implants. In *Proceedings of the 2003 IEEE International Symposium on Intelligent Control*, pages 63–68, Houston, Texas, Oct. 2003.

[213] S. Islam and A. Tayebi. New adaptive iterative learning control (AILC) for uncertain robot manipulators. In *Proceedings of the Canadian Conference on Electrical and Computer Engineering, 2004*, pages 1645–1651 Vol. 3, May 2004.

[214] A. K. Jain. *Fundamentals of Digital Image Processing*. Prentice Hall, Upper Saddle River, NJ, 1989.

[215] Luc Jaulin, Michel Kieffer, Olivier Didrit, and Éric Walter. *Applied Interval Analysis*. Springer, 2001.

[216] G. M. Jeong and C. H. Choi. Iterative learning control for linear discrete time nonminimum phase systems. *Automatica*, 38(2):287–291, 2002.

[217] Gu-Min Jeong and Chong-Ho Choi. Iterative learning control with advanced output data. In *Proceedings of the 3rd Asian Control Conference*, pages 1895–1900, Shanghai, China, 2000. ASCC.

[218] Gu-Min Jeong and Chong-Ho Choi. Iterative learning control with advanced output data for nonminimum phase systems. In *Proceedings of the 2001 American Control Conference*, pages 890–895, Arlington, VA, June 2001.

[219] Gu-Min Jeong and Chong-Ho Choi. Iterative learning control with advanced output data. *Asian Journal of Control*, 4(1):30–37, 2002.

[220] C. I. Jiang. Sufficient conditions for the asymptotic stability of interval matrices. *Int. J. Control*, 46:1803–1810, 1987.

[221] P. Jiang and R. Unbehauen. Iterative learning neural network control for nonlinear system trajectory tracking. *Neurocomputing*, 48:141–153, 2002.

[222] P. Jiang and R. Unbehauen. Robot visual servoing with iterative learning control. *IEEE Trans. on Systems Man and Cybernetics Part A – Systems and Humans*, 32(2):281–287, 2002.

[223] P. Jiang, P. Y. Woo, and R. Unbehauen. Iterative learning control for manipulator trajectory tracking without any control singularity. *Robotica*, 20:149–158 Part 2, 2002.

[224] Ping Jiang, Hui-Tang Chen, and Yue-Juan Wang. Iterative learning control for glass cutting robot. In *Proceedings of the 14th World Congress of IFAC*, pages 263–268, Beijing, China, 1999. IFAC.

[225] Ping Jiang and Huitang Chen. Robot teach by showing with iterative learning control. In *Proceedings of the 3rd Asian Control Conference*, Shanghai, China, 2000. ASCC.

[226] Ping Jiang and YangQuan Chen. Repetitive robot visual servoing via segmented trained neural network controller. In *Proceedings 2001 IEEE International Symposium on Computational Intelligence in Robotics and Automation*, pages 260–265, 29 July–1 Aug. 2001.

[227] Ping Jiang and YangQuan Chen. Singularity-free neural network controller with iterative training. In *Proceedings of the 2002 IEEE International Symposium on Intelligent Control*, pages 31–36, Vancouver, Canada, Oct. 2002.

[228] Ping Jiang and R. Unbehauen. An iterative learning control scheme with deadzone. In *Proceedings of the 38th IEEE Conference on Decision and Control*, pages 3816–3817, Phoenix, AZ, Dec. 1999.

[229] Ping Jiang, R. Unbehauen, and Peng-Yung Woo. Singularity-free indirect iterative learning control. In *Proceedings of the 40th IEEE Conference on Decision and Control*, pages 4903–4908, Orlando, FL, Dec. 2001.

[230] Daniel M. Joffe and Donald C. Panek, Jr. DTMF signaling on four-wire switched 56 KBPS lines. United States Patent 5,748,637, May 1998.

[231] Ramesh Johari and David Kium Hong Tan. End-to-end congestion control for the Internet: Delays and stability. *IEEE/ACM Trans. Networking*, 9:818–832, 2001.

[232] F. Kamal and M. Dahleh. Robust stability of multivariable interval control systems. *Int. J. Control*, 64(5):807–828, 1996.

[233] W. C. Karl and G. C. Verghese. A sufficient condition for the stability of interval matrix polynomials. *IEEE Trans. on Automatic Control*, 38(7):1139–1143, 1993.

[234] S. Kawamura and N. Sakagami. Analysis on dynamics of underwater robot manipulators based on iterative learning control and time-scale transformation. In *Proceedings of the IEEE International Conference on Robotics and Automation*, pages 1088–1094, Washington, DC, May 2002.

[235] V. L. Kharitonov. Asymptotic stability of an equilibrium position of a family of systems of linear differential equations. *Differential Equations*, 14:1483–1485, 1979.

[236] B. K. Kim, W. K. Chung, and Y. Youm. Robust learning control for robot manipulators based on disturbance observer. In *IECON Proceedings (Industrial Electronics Conference)*, volume 2, pages 1276–1282, Taipei, Taiwan, 1996.

[237] W. C. Kim, I. S. Chin, K. S. Lee, and J. H. Choi. Analysis and reduced-order design of quadratic criterion-based iterative learning control using singular value decomposition. *Computers & Chemical Engineering*, 24(8):1815–1819, 2000.

[238] Y. H. Kim and I. J. Ha. Asymptotic state tracking in a class of nonlinear systems via learning-based inversion. *IEEE Trans. on Automatic Control*, 45(11):2011–2027, 2001.

[239] Yong-Tae Kim, Heyoung Lee, Heung-Sik Noh, and Z. Bien. Robust higher-order iterative learning control for a class of nonlinear discrete-time systems. In *Proceedings of the IEEE International Conference on Systems, Man and Cybernetics*, pages 2219–2224, Oct. 2003.

[240] Koji Kinoshita, Takuya Sogo, and Norihiko Adachi. Adjoint-type iterative learning control for a single-link flexible arm. In *Proceedings of IFAC 15th World Congress*, Barcelona, Spain, July 2002. IFAC.

[241] Koji Kinosita, Takuya Sogo, and Norihiko Adachi. Iterative learning control using adjoint systems and stable inversion. *Asian Journal of Control*, 4(1):60–67, 2002.

[242] Kouji Kinosita, Takuya Sogo, and Norihiko Adachi. Iterative learning control using noncausal updating for non-minimum phase systems. In *Proceedings of the 3rd Asian Control Conference*, Shanghai, China, 2000. ASCC.

[243] H. Kokame and T. Mori. A Kharitonov-like theorem for interval polynomial matrices. *System and Control Letters*, 16:107–116, 1991.

[244] Olga Kosheleva, Vladik Kreinovich, Günther Mayer, and Hung T. Nguyen. Computing the cube of an interval matrix is NP-hard. Techni-

cal Report UTEP-CS-04-28, Univ. of Texas El Paso, Computer Science Department, 2004.

[245] Vladimir Kucera. *Analysis and Design of Discrete Linear Control Systems*. Prentice Hall, 1991.

[246] J. E. Kurek. Counterexample to iterative learning control of linear discrete-time multivariable systems. *Automatica*, 36(2):327–328, 2000.

[247] Huibert Kwakernaak and Michael Šebek. Polynomial *J*-spectral factorization. *IEEE Trans. on Automatic Control*, 39(2):315–328, 1994.

[248] S. I. Kwon, A. Regan, and Y. M. Wang. SNS superconducting RF cavity modeling – iterative learning control. *Nuclear Instruments & Methods in Physics Research Section A – Accelerators Spectrometers Dectectors and Associated Equipment*, 482(1–2):12–31, 2002.

[249] B. H. Lam, S. K. Panda, and J. X. Xu. Reduction of periodic speed ripples in PM synchronous motors using iterative learning control. In *Proceedings of the 26th Annual Confjerence of the IEEE Industrial Electronics Society*, pages 1406–1411, Nagoya Japan, 2000.

[250] B. H. Lam, S. K. Panda, J. X. Xu, and K. W. Lim. Torque ripple minimization in PM synchronous motor using iterative learning control. In *Proceedings of the 25th Annual Conference of the IEEE Industrial Electronics Society*, pages 1458–1463, San Jose, CA, Nov. 29–Dec. 3 1999.

[251] H. S. Lee and Z. Bien. Design issues on robustness and convergence of iterative learning controller. *Intelligent Automation and Soft Computing*, 8(2):95–106, 2002.

[252] J. H. Lee, S. Natarajan, and K. S. Lee. A model-based predictive control approach to repetitive control of continuous processes with periodic operations. *J. of Process Control*, 11(2):195–207, 2001.

[253] J. H. Lee, K. S. Lee, and W. C. Kim. Model-based iterative learning control with a quadratic criterion for time-varying linear systems. *Automatica*, 36(5):641–657, 2000.

[254] Jinho Lee, Insik Chin, Kwang S. Lee, and Jinhoon Choi. Temperature uniformity control in rapid thermal processing using an iterative learning control technique. In *Proceedings of the 3rd Asian Control Conference*, pages 419–423, Shanghai, China, 2000. ASCC.

[255] K. S. Lee, J. Lee, I. Chin, J. Choi, and J. H. Lee. Control of wafer temperature uniformity in rapid thermal processing using an optimal iterative learning control technique. *Industrial & Engineering Chemistry Research*, 40(7):1661–1672, 2001.

[256] K. S. Lee and J. H. Lee. Iterative learning control-based batch process control technique for integrated control of end product properties and transient profiles of process variables. *Journal of Process Control*, 13(7):607–621, 2003.

[257] K. S. Lee and J. H. Lee. Convergence of constrained model-based predictive control for batch processes. *IEEE Trans. on Automatic Control*, 45(10):1928–1932, 2000.

[258] Kwang S. Lee, Hyojin Ahn, In Sik Chin, Jay H. Lee, and Dae R. Yang. Optimal iterative learning control of wafer temperature uniformity in rapid thermal processing. In *Proceedings of IFAC 15th World Congress*, Barcelona, Spain, July 2002. IFAC.

[259] Kwang Soon Lee and Jay H. Lee. Constrained model-based predictive control combined with iterative learning for batch or repetitive processes. In *Proceedings of the 2nd Asian Control Conference*, Seoul, Korea, July 22–25 1997. ASCC.

[260] S. C. Lee, R. W. Longman, and M. Q. Phan. Direct model reference learning and repetitive control. *Intelligent Automation and Soft Computing*, 8(2):143–161, 2002.

[261] T. H. Lee, K. K. Tan, S. N. Huang, and H. F. Dou. Intelligent control of precision linear actuators. *Engineering Applications of Artificial Intelligence*, 13(6):671–684, 2001.

[262] T. H. Lee, K. K. Tan, S. Y. Lim, and H. F. Dou. Iterative learning control of permanent magnet linear motor with relay automatic tuning. *Mechatronics*, 10(1-2):169–190, 2000.

[263] Peter LeVoci. *Prediction final error level in learning and repetitive control*. PhD thesis, Columbia University, New York, NY, 2004.

[264] S. C. Li, X. H. Xu, and L. Ping. Feedback-assisted iterative learning control for batch polymerization reactor. *Advances in Neural Networks – ISNN 2004, Part 2 Lecutre Notes in Computer Science 3174*, pages 181–187, 2004.

[265] Shuchen Li, Xinhe Xu, and Ping Li. Frequency domain analysis for feedback-assisted iterative learning control. In *Proceedings of the International Conference on Information Acquisition, 2004*, pages 1–4, June 2004.

[266] W. Li, P. Maisser, and H. Enge. Self-learning control applied to vibration control of a rotating spindle by piezopusher bearings. *Proceedings of The Institution of Mechanical Engineers Part I – Journal of Systems and Control Engineering*, 218(13):185–196, 2004.

[267] Z. G. Li, C. W. Wen, Y. C. Soh, and Y. Q. Chen. Iterative learning control of linear time varying uncertain systems. In *Proceedings of the 3rd Asian Control Conference*, pages 1890–1894, Shanghai, China, 2000. ASCC.

[268] Y.-J. Liang and D. P. Looze. Performance and robustness issues in iterative learning control. In *Proceedings of 32nd IEEE Conference on Decision and Control*, pages 1990–1995, San Antonio, TX, December 1993.

[269] D. J. N. Limebeer, M. Green, and D. Walker. Discrete time H_∞ control. In *Proceedings of the 28th Conference on Decision and Control*, pages 392–396, Tampa, Florida, December 1989. IEEE.

[270] Z. Lin, L. Xu, and H. Fan. On minor prime factorization for n-D polynomial matrices. *IEEE Trans. Circuits and Systems II: Express Briefs*, 52(9):568–571, 2005.

[271] Qiang Ling and Michael D. Lemmon. Robust performance of soft real-time networked control systems with data dropouts. In *Proc. of the IEEE Conference on Decision and Control*, pages 1225–1230, Las Vegas, NV, 2002.

[272] Qiang Ling and Michael D. Lemmon. Soft real-time scheduling of networked control systems with dropouts governed by a Markov chain. In *Proc. of the American Control Conference*, pages 4845–4850, Denver, CO, 2003.

[273] Qiang Ling and Michael D. Lemmon. Power spectral analysis of networked control systems with data dropouts. *IEEE Trans. Automatic Control*, 49:955–960, 2004.

[274] David Liu, Li-Chen Fu, Su-Hau Hsu, and Teng-Kai Kuo. Analysis on an on-line iterative correction control law for visual tracking. In *Proceedings of the 3rd Asian Control Conference*, Singapore, Sept. 2002. ASCC.

[275] Hui Liu and Danwei Wang. Convergence of CMAC network learning control for a class of nonlinear dynamic system. In *Proceedings of the 1999 IEEE International Symposium on Intelligent Control/Intelligent Systems and Semiotics*, pages 108–113, Cambridge, MA, Sept. 1999.

[276] Chun Ping Lo. *Departure angle based and frequency based compensator design in learning and repetitive control*. PhD thesis, Columbia University, New York, NY, 2004.

[277] R. W. Longman and Y. C. Huang. The phenomenon of apparent convergence followed by divergence in learning and repetitive control. *Intelligent Automation and Soft Computing*, 8(2):107–128, 2002.

[278] R. W. Longman and S. L. Wirkander. Automated tuning concepts for iterative learning and repetitive control laws. In *Proceedings of the 37th IEEE Conference on Decision and Control*, pages 192–198, Tampa, FL, Dec. 1998.

[279] Richard W. Longman. Iterative learning control and repetitive control for engineering practice. *Int. J. of Control*, 73(10):930–954, 2000.

[280] Yonggui Lv and Yanding Wei. Study on open-loop precision positioning control of a micropositioning platform using a piezoelectric actuator. In *Proceedings of the Fifth World Congress on Intelligent Control and Automation*, pages 1255–1259, Hangzhou, China, June 2004.

[281] P. M. Lynch, J. F. Figueroa, and J. DePaso. A prototype intelligent control structure using intermittent multiple independent measurements. In *Proc. of the IEEE 1990 Southeastcon*, 1990.

[282] Lili Ma. *Vision-based measurements for dynamic systems and control*. PhD thesis, Utah State University, 2004.

[283] Magdi S. Mahmoud and Abdulla Ismail. Role of delays in networked control systems. In *Proc. of the ICECS*, 2003.

[284] K. Mainali, S. K. Panda, J. X. Xu, and T. Senjyu. Position tracking performance enhancement of linear ultrasonic motor using iterative learning control. In *Proceedings of the IEEE 35th Annual Power Elec-*

tronics Specialists Conference, pages 4844–4849, Aachen, Germany, June 2004. PESC 04.

[285] K. Mainali, S. K. Panda, J. X. Xu, and T. Senjyu. Repetitive position tracking performance enhancement of linear ultrasonic motor with sliding mode-cum-iterative learning control. In *Proceedings of the IEEE International Conference on Mechatronics, 2004*, pages 352–357, June 2004.

[286] O. Markusson, H. Hjalmarsson, and M. Norrlöf. Iterative learning control of nonlinear non-minimum phase systems and its application to system and model inversion. In *Proceedings of the 40th IEEE Conference on Decision and Control*, pages 4481–4482, Orlando, FL, Dec. 2001.

[287] Ola Markusson. *Model and system inversion with applications in nonlinear system identification and control*. PhD thesis, Kungliga Tekniska Hogskolan (Sweden), 2002.

[288] Ola Markusson, Håkan Hjalmarsson, and Mikael Norrlöf. A general framework for iterative learning control. In *Proceedings of IFAC 15th World Congress*, Barcelona, Spain, July 2002. IFAC.

[289] M. Matsushima, T. Hashimoto, and F. Miyazaki. Learning the robot table tennis task – ball control and rally with a human. In *Proceedings of the IEEE International Conference on Systems, Man and Cybernetics*, pages 2962–2969, Oct. 2003.

[290] Günter Mayer. On the convergence of powers of interval matrices. *Linear Algebra and Its Applications*, 58:201–216, 1984.

[291] Leonard Meirovitch. *Computational Methods in Structural Dynamics*. Sijthoff & Noordhoff, Rockville, Maryland, 1980.

[292] M. Mezghani, M. V. Lelann, G. Roux, M. Cabassud, B. Dahhou, and G. Casmatta. Experimental application of the iterative learning control to the temperature control of batch reactor. In *Proceedings of IFAC 15th World Congress*, Barcelona, Spain, July 2002. IFAC.

[293] M. Mezghani, G. Roux, M. Cabassud, B. Dahhou, M. V. Le Lann, and G. Casamatta. Robust iterative learning control of an exothermic semi-batch chemical reactor. *Mathematics and Computers in Simulation*, 57(6):367–385, 2001.

[294] M. Mezghani, G. Roux, M. Cabassud, M. W. Le Lann, B. Dahhou, and G. Casamatta. Application of iterative learning control to an exothermic semibatch chemical reactor. *IEEE Trans. on Control Systems Technology*, 10(6):822–834, 2002.

[295] Y. Miyasato. Iterative learning control of robotic manipulators by hybrid adaptation schemes. In *Proceedings of the 42nd IEEE Conference on Decision and Control*, pages 4428–4433, Maui, Hawaii, Dec. 2003.

[296] J. H. Moon, T. Y. Doh, and M. J. Chung. A robust approach to iterative learning control design for uncertain systems. *Automatica*, 34(8):1001–1004, 1998.

[297] Todd K. Moon and Wynn C. Stirling. *Mathematical Methods and Algorithms for Signal Processing*. Prentice Hall, 1999.

[298] K. L. Moore. *Iterative Learning Control for Deterministic Systems.* Advances in Industrial Control. Springer-Verlag, 1993.

[299] K. L. Moore. Iterative learning control – an expository overview. *Applied & Computational Controls, Signal Processing, and Circuits*, 1(1):151–241, 1999.

[300] K. L. Moore. An iterative learning control algorithm for systems with measurement noise. In *Proc. of the 38th IEEE Conference on Decision and Control*, pages 270–275, Phoenix, Arizona, Dec. 1999. IEEE.

[301] K. L. Moore. A matrix fraction approach to higher-order iterative learning control: 2-D dynamics through repetition-domain filtering. In *Proceedings of the Second International Workshop on Mutidimensional (nD) Systems*, pages 99–104, Czocha Castle, Lower Silesia, Poland, June 2000.

[302] K. L. Moore. On the relationship between iterative learning control and one-step ahead minimum prediction error control. In *Proc. of the Asian Control Conference*, pages 1861–1865, Shanghai, China, July 2000.

[303] K. L. Moore. An observation about monotonic convergence in discrete-time, P-type iterative learning control. In *Proceedings of the 2001 IEEE International Symposium on Intelligent Control*, pages 45–49, Mexico City, Mexico, Sept. 2001.

[304] K. L. Moore and V. Bahl. Iterative learning control for multivariable systems with an application to mobile robot path tracking. In *Proceedings of the International Conference on Automation, Robotics, and Control*, pages 396–401, Singapore, Dec. 2000.

[305] K. L. Moore and Y. Q. Chen. A separative high-order framework for monotonic convergent iterative learning controller design. In *Proceedings of the 2003 American Control Conference*, pages 3644–3649, Denver, CO, June 4–6 2003.

[306] K. L. Moore and YangQuan Chen. On monotonic convergence of high order iterative learning update laws. In *Invited Session on High-order Iterative Learning Control* at the 15th IFAC Congress, pages 1–6, Barcelona, Spain, July 21–26 2002. IFAC.

[307] K. L. Moore, M. Dahleh, and S. P. Bhattacharyya. Iterative learning control: a survey and new results. *J. of Robotic Systems*, 9(5):563–594, 1992.

[308] Kevin L. Moore. Multi-loop control approach to designing iterative learning controllers. In *Proceedings of the 37th IEEE Conference on Decision and Control*, pages 666–671, Tampa, Florida, 1998.

[309] Kevin L. Moore. A non-standard iterative learning control approach to tracking periodic signals in discrete-time non-linear systems. *Int. J. of Control*, 73(10):955–967, 2000.

[310] Kevin L. Moore, YangQuan Chen, and Vikas Bahl. Feedback controller design to ensure monotonic convergence in discrete-time, P-type iterative learning control. In *Proceedings of the 2002 Asian Control Conference*, pages 440–445, Singapore, Sept. 2002.

[311] Kevin L. Moore, YangQuan Chen, and Vikas Bahl. Monotonically convergent iterative learning control for linear discrete-time systems. *Automatica*, 41(9):1529–1537, 2005.

[312] Kevin L. Moore and Jian-Xin Xu (Eds). *Special Issue:* Iterative Learning Control. *Int. J. of Control*, 73(10):819–999, 2000.

[313] K.L. Moore. An iterative learning control algorithm for systems with measurement noise. In *Proceedings of the 38th IEEE Conference on Decision and Control*, pages 270–275, Phoenix, AZ, Dec. 1999.

[314] R. E. Moore. *Interval Analysis*. Prentice-Hall, Englewood Cliffs, NJ, 1966.

[315] Pham Thuc Anh Nguyen and Suguru Arimoto. Learning motion of dexterous manipulation for a pair of multi-DOF fingers with soft-tips. *Asian Journal of Control*, 4(1):11–20, 2002.

[316] Pham Thuc Anh Nguyen, Hyun-Yong Han, S. Arimoto, and S. Kawamura. Iterative learning of impedance control. In *Proceedings of the 1999 IEEE/RSJ International Conference on Intelligent Robots and Systems*, pages 653–658, Kyongju, South Korea, Oct. 1999.

[317] G. Nijsse, M. Verhaegen, and N. J. Doelman. A new subspace based approach to iterative learning control. In *Proceedings of the 2001 European Control Conference*, Seminário de Vilar, Porto, Portugal, Sept. 2001. ECC.

[318] A. Nishiki, T. Sogo, and N. Adachi. Tip position control of a one-link flexible arm by adjoint-type iterative learning control. In *Proceedings of the 41st SICE Annual Conference*, pages 1551–1555, Aug. 2002.

[319] Ji-Ho Noh, Hyun-Sik Ahn, and Do-Hyun Kim. Hybrid position/force control of a two-link SCARA robot using a fuzzy-tuning repetitive controller. In *Proceedings of the 3rd Asian Control Conference*, Shanghai, China, 2000. ASCC.

[320] M. Norrlöf. Comparative study on first and second order ILC-frequency domain analysis and experiments. In *Proceedings of the 39th IEEE Conference on Decision and Control*, pages 3415–3420, Sydney, NSW, Dec. 2000.

[321] M. Norrlöf. *Iterative Learning Control: Analysis, Design, and Experiments*. PhD thesis, Linköpings Universitet, Linköping, Sweden, 2000.

[322] M. Norrlöf. An adaptive approach to iterative learning control with experiments on an industrial robot. In *Proceedings of the 2001 European Control Conference*, Seminário de Vilar, Porto, Portugal, Sept. 2001. ECC.

[323] M. Norrlöf. An adaptive iterative learning control algorithm with experiments on an industrial robot. *IEEE Trans. on Robotics and Automation*, 18(2):245–251, 2002.

[324] M. Norrlöf. Iteration varying filters in iterative learning control. In *Proceedings of the 3rd Asian Control Conference*, Singapore, Sept. 2002. ASCC.

[325] M. Norrlöf. Disturbance rejection using an ILC algorithm with iteration varying filters. *Asian J. of Control*, 6(3):432–438, 2004.

[326] M. Norrlöf and S. Gunnarsson. A frequency domain analysis of a second order iterative learning control algorithm. In *Proceedings of the 38th IEEE Conference on Decision and Control*, pages 1587–1592, Phoenix, AZ, Dec. 1999.

[327] M. Norrlöf and S. Gunnarsson. Disturbance aspects of iterative learning control. *Engineering Applications of Artificial Intelligence*, 14(1):87–94, 2001.

[328] M. Norrlöf and S. Gunnarsson. Disturbance aspects of high order iterative learning control. In *Proceedings of IFAC 15th World Congress*, Barcelona, Spain, July 2002. IFAC.

[329] M. Norrlöf and S. Gunnarsson. Experimental comparison of some classical iterative learning control algorithms. *IEEE Trans. on Robotics and Automation*, 18(4):636–641, 2002.

[330] M. Norrlöf and S. Gunnarsson. Some new results on current iteration tracking error ILC. In *Proceedings of the 3rd Asian Control Conference*, Singapore, Sept. 2002. ASCC.

[331] M. Norrlöf and S. Gunnarsson. Time and frequency domain convergence properties in iterative learning control. *Int. J. of Control*, 75(14):1114–1126, 2002.

[332] Mikael Norrlöf. *Iterative learning control: analysis, design, and experiments*. PhD thesis, Linköping Studies in Science and Technology (Sweden), 2000.

[333] R. Ogoshi, T. Sogo, and N. Adachi. Adjoint-type iterative learning control for nonlinear nonminimum phase system – application to a planar model of a helicopter. In *Proceedings of the 41st SICE Annual Conference*, pages 1547 –1550, Shanghai, China, Aug. 2002.

[334] Sang June Oh. *Synthesis and analysis of design methods for improved tracking performance in iterative learning and repetitive control*. PhD thesis, Columbia University, New York, 2004.

[335] Manuel Olivares, Pedro Albertos, and Antonio Sala. Iterative learning controller design for multivariable systems. In *Proceedings of IFAC 15th World Congress*, Barcelona, Spain, July 2002. IFAC.

[336] Alexander Olshevsky and Vadim Olshevsky. The Kharitonov theorem and Bezoutians. In *Sixteenth International Symposium on Mathematical Theory of Networks and Systems (MTNS2004)*, Katholieke Universiteit Leuven, Belgium, July 2004.

[337] Giuseppe Oriolo, Stefano Panzieri, and Giovanni Ulivi. Learning optimal trajectories for non-holonomic systems. *Int. J. of Control*, 73(10):980–991, 2000.

[338] J. M. Ortega and W. C. Rheinboldt. *Iterative Solution of Nonlinear Equations in Several Variables*. Academic Press, New York, 1970.

[339] Jamahl W. Overstreet and Anthony Tzes. An Internet-based real-time control engineering laboratory. *IEEE Control Systems Magazine*, 19:19–34, 1999.

[340] D. H. Owens and K. Feng. Parameter optimization in iterative learning control. *Int. J. of Control*, 76(11):1059–1069, 2003.

[341] D. H. Owens and J. Hätönen. Convex modifications to an iterative learning control law. In *Proceedings of the 15th IFAC World Congress on Automatic Control*, Barcelona, Spain, July 2002. IFAC.

[342] D. H. Owens and J. J. Hätönen. A new optimality based adaptive ILC-algorithm. In *Proceedings of the 7th International Conference on Control, Automation, Robotics and Vision*, pages 1496–1501, Singapore, Dec. 2002.

[343] D. H. Owens and G. Munde. Error convergence in an adaptive iterative learning controller. *Int. J. of Control*, 73(10):851–857, 2000.

[344] D. H. Owens and E. Rogers. Iterative learning control-recent progress and open research problems. In *Learning Systems for Control, IEE Seminar*, pages 7/1–7/3, Birmingham, U.K., May 2000.

[345] D. H. Owens and E. Rogers. Comments on "On the equivalence of causal LTI iterative learning control and feedback control" by P. B. Goldsmith. *Automatica*, 40(5):895–898, 2004.

[346] D. H. Owens, E. Rogers, and K. L. Moore. Analysis of linear iterative learning control schemes using repetitive process theory. *Asian Journal of Control*, 4(1):68–89, 2002.

[347] D. H. Owens. The benefits of prediction in learning control algorithms. In *IEE Two-Day Workshop on Model Predictive Control: Techniques and Applications – Day 1*, pages 3/1–3/3, London, U.K., April 1999.

[348] D. H. Owens, N. Amann, E. Rogers, and M. French. Analysis of linear iterative learning control schemes – A 2D systems/repetitive processes approach. *Multidimensional Systems and Signal Processing*, 11(1–2):125–177, 2000.

[349] D. H. Owens and G. Munde. Error convergence in an adaptive iterative learning controller. *Int. J. of Control*, 73(10):851–857, 2000.

[350] D. H. Owens and G.S. Munde. Universal adaptive iterative learning control. In *Proceedings of the 37th IEEE Conference on Decision and Control*, pages 181–185, Tampa, FL, Dec. 1998.

[351] J. J. Hatonen, D. H. Owens, and K. L. Moore. An algebraic approach to iterative learning control. *Int. J. Control*, 77(1):45–54, 2004.

[352] M. Pandit and K. H. Buchheit. Optimizing iterative learning control of cyclic production processes with application to extruders. *IEEE Trans. on Control Systems Technology*, 7(3):382–390, 1999.

[353] K. H. Park and Z. Bien. A generalized iterative learning controller against initial state error. *Int. J. of Control*, 73(10):871–881, 2000.

[354] K. H. Park, Z. Bien, and D. H. Hwang. Design of an iterative learning controller for a class of linear dynamic systems with time delay. *IEE Proceedings – Control Theory and Applications*, 145(6):507–512, 1998.

[355] K. H. Park, Z. Bien, and D. H. Hwang. A study on the robustness of a PID-type iterative learning controller against initial state error. *Int. J. of Systems Science*, 30(1):49–59, 1999.

[356] Kwang-Hyun Park and Zeungnam Bien. A generalized iterative learning controller against initial state error. *Int. J. of Control*, 73(10):871–881, 2000.

[357] Kwang-Hyun Park and Zeungnam Bien. Intervalized iterative learning control for monotone convergence in the sense of sup-norm. In *Proceedings of the 3rd Asian Control Conference*, pages 2899–2903, Shanghai, China, 2000. ASCC.

[358] Kwang-Hyun Park and Zeungnam Bien. A study on iterative learning control with adjustment of learning interval for monotone convergence in the sense of sup-norm. *Asian Journal of Control*, 4(1):111–118, 2002.

[359] Kwang-Hyun Park and Zeungnam Bien. A study on robustness of iterative learning controller with input saturation against time-delay. In *Proceedings of the 3rd Asian Control Conference*, Singapore, Sept. 2002. ASCC.

[360] W. Paszke, K. Galkowski, E. Rogers, and D. H. Owens. H_∞ control of discrete linear repetitive processes. In *Proceedings of the 42nd IEEE Conference on Decision and Control*, pages 628–633, Maui, Hawaii, Dec. 2003.

[361] K. S. Peterson and A. G. Stefanopoulou. Extremum seeking control for soft landing of an electromechanical valve actuator. *Automatica*, 40(6):1063–1069, 2004.

[362] Djordjija B. Petkovski. Stability analysis of interval matrices: Improved bounds. *Int. J. Control*, 48(6):2265–2273, 1988.

[363] M. Q. Phan, R. W. Longman, and K. L. Moore. Unified formulation of linear iterative learning control. In *AAS/AIAA Space Flight Mechanics Meeting*, pages AAA 00–106, Clearwater, Florida, Jan. 2000.

[364] M. Q. Phan and H. Rabitz. A self-guided algorithm for learning control of quantum-mechanical systems. *Journal of Chemical Physics*, 110(1):34–41, 1999.

[365] Minh Q. Phan and Richard W. Longman. Higher-order iterative learning control by pole placement and noise filtering. In *Proceedings of IFAC 15th World Congress*, Barcelona, Spain, July 2002. IFAC.

[366] M. Q. Phan and J. A. Frueh. Model reference adaptive learning control with basis functions. In *Proceedings of the 38th IEEE Conference on Decision and Control*, pages 251–257, Phoenix, AZ, Dec. 1999.

[367] Nonglak Phetkong. *Learning control and repetitive control of a high speed nonlinear cam follower system*. PhD thesis, Lehigh University, Bethehem, PA, 2002.

[368] Dao-Ying Pi and K. Panaliappan. Robustness of discrete nonlinear systems with open-closed-loop iterative learning control. In *Proceedings of the 2002 International Conference on Machine Learning and Cybernetics*, pages 1263–1266, Nov. 2002.

[369] Daoying Pi, D. Seborg, Jianxia Shou, Youxian Sun, and Qing Lin. Analysis of current cycle error assisted iterative learning control for discrete nonlinear time-varying systems. In *IEEE International Conference on Systems, Man, and Cybernetics*, pages 3508–3513, Nashville, TN, Oct. 2000.

[370] Yang-Ming Pok, Kok-Hwa Liew, and Jian-Xin Xu. Fuzzy PD iterative learning control algorithm for improving tracking accuracy. In *1998 IEEE International Conference on Systems, Man, and Cybernetics*, pages 1603–1608, San Diego, CA, Oct. 1998.

[371] Joe Paul Predina and Harold L. Broberg. Tuned open-loop switched to closed-loop method for rapid point-to-point movement of a periodic motion control system. United States Patent 6,686,716, February 2004.

[372] P. J. Psarrakos and M. J. Tsatsomeros. On the stability radius of matrix polynomials. *Linear and Multilinear Algebra*, 50:151–165, 2002.

[373] W. Z. Qian, S. K. Panda, and J. X. Xu. Torque ripple minimization in PM synchronous motors using iterative learning control. *IEEE Trans. on Power Electronics*, 19(2):272–279, 2004.

[374] Weizhe Qian, S. K. Panda, and J. X. Xu. Periodic speed ripples minimization in PM synchronous motors using repetitive learning variable structure control. In *Proceedings of the 3rd Asian Control Conference*, Singapore, 2002. ASCC.

[375] Weizhe Qian, S. K. Panda, and J. X. Xu. Torque ripples reduction in PM synchronous motor using frequency-domain iterative learning control. In *Proceedings of the Fifth International Conference on Power Electronics and Drive Systems*, pages 1636–1641, Nov. 2003.

[376] Wubi Qin and Lilong Cai. A Fourier series based iterative learning control for nonlinear uncertain systems. In *Proceedings of the 2001 IEEE/ASME International Conference on Advanced Intelligent Mechatronics*, pages 482–487, July 2001.

[377] Wubi Qin and Lilong Cai. A frequency domain iterative learning control for low bandwidth system. In *Proceedings of the 2001 American Control Conference*, pages 1262–1267, Arlington, VA, June 2001.

[378] Z. Qu and Jianxin Xu. Learning unknown functions in cascaded nonlinear systems. In *Proceedings of the 37th IEEE Conference on Decision and Control*, pages 165–169, Tampa, FL, Dec. 1998.

[379] Z. H. Qu. An iterative learning algorithm for boundary control of a stretched moving string. *Automatica*, 38(5):821–827, 2002.

[380] Z. H. Qu and J. X. Xu. Asymptotic learning control for a class of cascaded nonlinear uncertain systems. *IEEE Trans. on Automatic Control*, 47(8):1369–1376, 2002.

[381] Zhihua Qu. An iterative learning algorithm for boundary control of a stretched string on a moving transporter. In *Proceedings of the 3rd Asian Control Conference*, Shanghai, China, 2000. ASCC.

[382] Zhihua Qu and Jian-Xin Xu. Model-based learning controls and their comparisons using Lyapunov direct method. *Asian Journal of Control*, 4(1):99–110, 2002.

[383] A. Robertsson, D. Scalamogna, M. Grundelius, and R. Johansson. Cascaded iterative learning control for improved task execution of optimal control. In *Proceedings of the IEEE International Conference on Robotics and Automation*, pages 1290–1295, Washington DC, May 2002.

[384] J. A. Rogas and J. M. Collado. Stability of interval matrices using the distance to the set of unstable matrices. In *Proceedings of the 1994 American Control Conference*, Baltimore, MD, 1994.

[385] E. Rogers, K. Galkowski, A. Gramacki, J. Gramacki, and D. H. Owens. Stability and controllability of a class of 2-D linear systems with dynamic boundary conditions. *IEEE Trans. on Circuits and Systems I – Fundamental Theory and Applications*, 49(2):181–195, 2002.

[386] E. Rogers, J. Lam, K. Galkowski, S. Xu, J. Wood, and D. H. Owens. LMI based stability analysis and controller design for a class of 2D discrete linear systems. In *Proceedings of the 40th IEEE Conference on Decision and Control*, pages 4457–4462, Orlando, FL, Dec. 2001.

[387] E. Rogers and D. H. Owens. *Stability Analysis for Linear Repetitive Processes*. Springer-Verlag, Berlin, 1992.

[388] J. Rohn. Positive definiteness and stability of interval matrices. *SIAM J. Matrix Anal. Appl.*, 15(1):175–184, 1994.

[389] J. Rohn. An algorithm for checking stability of symmetric interval matrices. *IEEE Trans. on Automatic Control*, 41(1):133–136, 1996.

[390] Dick De Roover and Okko H. Bosgra. Synthesis of robust multivariable iterative learning controllers with application to a wafer stage motion system. *Int. J. of Control*, 73(10):968–979, 2000.

[391] Dick De Roover, Okko H. Bosgra, and Maarten Steinbuch. Internal-model-based design of repetitive and iterative learning controllers for linear multivariable systems. *Int. J. of Control*, 73(10):914–929, 2000.

[392] I. Rotariu, R. Ellenbroek, and M. Steinbuch. Time-frequency analysis of a motion system with learning control. In *Proceedings of the 2003 American Control Conference*, pages 3650–3654, Denver, CO, June 2003.

[393] Xiao-E Ruan, Bai-Wu Wan, and Hong-Xia Gao. The iterative learning control for saturated nonlinear industrial control systems with dead zone. In *Proceedings of the 3rd Asian Control Conference*, pages 1554–1557, Shanghai, China, 2000. ASCC.

[394] M. Rzewuski, M. French, E. Rogers, and D. H. Owens. Prediction in iterative learning control versus learning along the trials. In *Proceedings of IFAC 15th World Congress*, Barcelona, Spain, July 2002. IFAC.

[395] S. S. Saab. A discrete-time stochastic learning control algorithm. *IEEE Trans. on Automatic Control*, 46(6):877–887, 2001.

[396] S. S. Saab. On a discrete-time stochastic learning control algorithm. *IEEE Trans. on Automatic Control*, 46(8):1333–1336, 2001.

[397] S. S. Saab. Stochastic P-type/D-type iterative learning control algorithms. *Int. J. of Control*, 76(2):139–148, 2003.

[398] S. S. Saab. A stochastic iterative learning control algorithm with application to an induction motor. *Int. J. of Control*, 77(2):144–163, 2004.

[399] N. C. Sahoo, J. X. Xu, and S. K. Panda. Application of iterative learning for constant torque control of switched reluctance motors. In *Proceedings of the 14th World Congress of IFAC*, pages 47–52, Beijing, China, 1999. IFAC.

[400] N. C. Sahoo, J. X. Xu, and SK. Panda. Low torque ripple control of switched reluctance motors using iterative learning. *IEEE Trans. on Energy Conversion*, 16(4):318–326, 2001.

[401] S. K. Sahoo, S. K. Panda, and J. X. Xu. Iterative learning based torque controller for switched reluctance motors. In *Proceedings of the 29th Annual Conference of the IEEE Industrial Electronics Society*, pages 2459–2464, Nov. 2003.

[402] S. K. Sahoo, S. K. Panda, and J. X. Xu. Iterative learning control based direct instantaneous torque control of switched reluctance motors. In *Proceedings of the IEEE 35th Annual Power Electronics Specialists Conference, 2004*, pages 4832–4837. PESC 04, June 2004.

[103] N. Sakagami, M. Inoue, and S. Kawamura. Theoretical and experimental studies on iterative learning control for underwater robots. *Int. J. of Offshore and Polar Engineering*, 13(2):120–127, 2003.

[404] N. Sakagami and S. Kawamura. Time optimal control for underwater robot manipulators based on iterative learning control and time-scale transformation. In *Proceedings of OCEANS 2003*, pages 1180–1186, Sept. 2003.

[405] E. Schamiloglu, G. T. Park, V. S. Soualian, C. T. Abdallah, and F. Hegeler. Advances in the control of a smart tube high power backward wave oscillator. In *Proceedings of the 12th IEEE International Pulsed Power Conference*, pages 852–855, Monterey, CA, June 1999.

[406] W.G. Seo, B.H. Park, and J.S. Lee. Adaptive fuzzy learning control for a class of nonlinear dynamic systems. *Int. J. of Intelligent Systems*, 15(12):1157–1175, 2000.

[407] Zhong-Ke Shi. Real-time learning control method and its application to AC-servomotor control. In *Proceedings of the 2002 International Conference on Machine Learning and Cybernetics*, pages 900–905, Beijing, China, Nov. 2002.

[408] D. M. Shin, J. Y. Choi, and J. S. Lee. A P-type iterative learning controller for uncertain robotic systems with exponentially decaying error bounds. *Journal of Robotic Systems*, 20(2):79–91, 2003.

[409] Jianxia Shou, Daoying Pi, and Wenhai Wang. Sufficient conditions for the convergence of open-closed-loop PID-type iterative learning control for nonlinear time-varying systems. In *Proceedings of the IEEE International Conference on Systems, Man and Cybernetics, 2003*, pages 2557–2562, Oct. 2003.

[410] Jianxia Shou, Zhengjiang Zhang, and Daoying Pi. On the convergence of open-closed-loop D-type iterative learning control for nonlinear systems. In *Proceedings of the IEEE International Symposium on Intelligent Control*, pages 963–967, Houston, Texas, Oct. 2003.

[411] B. Sinopoli, L. Schenato, M. Franceschetti, K. Poolla, M. I. Jordan, and S. S. Sastry. Kalman filtering with intermittent observations. *IEEE Trans. Automatic Control*, 49:1453–1464, 2004.

[412] S. Craig Smith and Peter Seiler. Estimation with lossy measurements: Jump estimators for jump systems. *IEEE Trans. Automatic Control*, 48:2163–2171, 2003.

[413] T. Sogo. Stable inversion for nonminimum phase sampled-data systems and its relation with the continuous-time counterpart. In *Proceedings of the 41st IEEE Conference on Decision and Control*, pages 3730–3735, Las Vegas, Nevada, Dec. 2002.

[414] T. Sogo, K. Kinoshita, and N. Adachi. Iterative learning control using adjoint systems for nonlinear non-minimum phase systems. In *Proceedings of the 39th IEEE Conference on Decision and Control*, pages 3445–3446, Sydney, NSW, Dec. 2000.

[415] Takuya Sogo and Norihiko Adachi. Convergence rates and robustness of iterative learning control. In *Proceedings of the 35th Conference on Decision and Control*, pages 3050–3055, Kobe, Japan, Dec. 1996. IEEE.

[416] Thuantong Songchon. *The waterbed effect and stability in learning/repetitive control*. PhD thesis, Columbia University, New York, NY, 2001.

[417] Szathys Songschon and R. W. Longman. Comparison of the stability boundary and the frequency response stability condition in learning and repetitive control. *Int. J. Appl. Math. Comput. Sci.*, 13(2):169–177, 2003.

[418] Te-Jen Su and Wen-Jye Shyr. Robust D-stability for linear uncertain discrete time-delay systems. *IEEE Trans. on Automatic Control*, 39(2):425–428, 1994.

[419] T. Sugie. On iterative learning control. In *Proceedings of the International Conference on Informatics Research for Development of Knowledge Society Infrastructure*, pages 214–220, March 2004.

[420] T. Sugie and K. Hamamoto. Iterative learning control – an identification oriented approach. In *Proceedings of the 41st SICE Annual Conference*, pages 2563–2566, Aug. 2002.

[421] B. Sulikowski, K. Galkowski, E. Rogers, and D. H. Owens. Output feedback control of discrete linear repetitive processes. *Automatica*, 40(12):2167–2173, 2004.

[422] D. Sun and J. K. Mills. Performance improvement of industrial robot trajectory tracking using adaptive-learning scheme. *Journal of Dynamic Systems Measurement and Control – Trans. of the ASME*, 121(2):285–292, 1999.

[423] Dong Sun and J. K. Mills. Adaptive learning control of robotic systems with model uncertainties. In *Proceedings of the IEEE International*

Conference on Robotics and Automation, pages 1847–1852, Leuven, May 1998.

[424] Dong Sun and J. K. Mills. High-accuracy trajectory tracking of industrial robot manipulator using adaptive-learning scheme. In *Proceedings of the 1999 American Control Conference*, pages 1935–1939, San Diego, CA, June 1999.

[425] M. X. Sun and D. W. Wang. Anticipatory iterative learning control for nonlinear systems with arbitrary relative degree. *IEEE Trans. on Automatic Control*, 46(5):783–788, 2001.

[426] M. X. Sun and D. W. Wang. Initial condition issues on iterative learning control for non-linear systems with time delay. *Int. J. of Systems Science*, 32(11):1365–1375, 2001.

[427] M. X. Sun and D. W. Wang. Iterative learning control design for uncertain dynamic systems with delayed states. *Dynamics and Control*, 10(4):341–357, 2001.

[428] M. X. Sun and D. W. Wang. Sampled-data iterative learning control for nonlinear systems with arbitrary relative degree. *Automatica*, 37(2):283–289, 2001.

[429] M. X. Sun and D. W. Wang. Closed-loop iterative learning control for non-linear systems with initial shifts. *Int. J. of Adaptive Control and Signal Processing*, 16(7):515–538, 2002.

[430] M. X. Sun and D. W. Wang. Iterative learning control with initial rectifying action. *Automatica*, 38(7):1177–1182, 2002.

[431] M. X. Sun and D. W. Wang. Initial shift issues on discrete-time iterative learning control with system relative degree. *IEEE Trans. on Automatic Control*, 48(1):144–148, 2003.

[432] M. X. Sun, D. W. Wang, and Y. Y. Wang. Sampled-data iterative learning control with well-defined relative degree. *Int. J. of Robust and Nonlinear Control*, 14(8):719–739, 2004.

[433] Mingxuan Sun and Danwei Wang. Sampled-data iterative learning control for a class of nonlinear systems. In *Proceedings of the 1999 IEEE International Symposium on Intelligent Control/Intelligent Systems and Semiotics*, pages 338–343, Cambridge, MA, Sept. 1999.

[434] Mingxuan Sun and Danwei Wang. Initial position shift problem and its ILC solution for nonlinear systems with a relative degree. In *Proceedings of the 3rd Asian Control Conference*, pages 1900B1–1900B2, Shanghai, China, 2000. ASCC.

[435] Mingxuan Sun and Danwei Wang. Robust discrete-time iterative learning control: initial shift problem. In *Proceedings of the 40th IEEE Conference on Decision and Control*, pages 1211–1216, Orlando, FL, Dec. 2001.

[436] Mingxuan Sun and Danwei Wang. Higher relative degree nonlinear systems with ILC using lower-order differentiations. *Asian Journal of Control*, 4(1):38–48, 2002.

[437] Mingxuan Sun and Danwei Wang. Higher relative degree nonlinear systems with sampled-data ILC using lower-order differentiations. In *Proceedings of the 3rd Asian Control Conference*, Singapore, Sept. 2002. ASCC.

[438] Mingxuan Sun, Danwei Wang, and Guangyan Xu. Initial shift problem and its ILC solution for nonlinear systems with higher relative degree. In *Proceedings of the 2000 American Control Conference*, pages 277–281, Chicago, IL, June 2000.

[439] Mingxuan Sun, Danwei Wang, and Guangyan Xu. Sampled-data iterative learning control for SISO nonlinear systems with arbitrary relative degree. In *Proceedings of the 2000 American Control Conference*, pages 667–671, Chicago, IL, June 2000.

[440] Peng Sun, Zhong Fang, and Zhengzhi Han. Sampled-data iterative learning control for singular systems. In *Proceedings of the 4th World Congress on Intelligent Control and Automation*, pages 555–559, Shanghai, China, June 2002.

[441] K. K. Tan, H. F. Dou, Y. Q. Chen, and T. H. Lee. High precision linear motor control via relay-tuning and iterative learning based on zero-phase filtering. *IEEE Trans. on Control Systems Technology*, 9(2):244–253, 2001.

[442] K. K. Tan, S. N. Huang, and T. H. Lee. Predictive iterative learning control. In *Proceedings of the 3rd Asian Control Conference*, Singapore, Sept. 2002. ASCC.

[443] K. K. Tan, S. N. Huang, T. H. Lee, and S. Y. Lim. A discrete-time iterative learning algorithm for linear time-varying systems. *Engineering Applications of Artificial Intelligence*, 16(3):185–190, 2003.

[444] K. K. Tan, S. Y. Lim, T. H. Lee, and H. F. Dou. High-precision control of linear actuators incorporating acceleration sensing. *Robotics and Computer-Integrated Manufacturing*, 16(5):295–305, 2000.

[445] K. K. Tan and J. C. Tang. Learning-enhanced PI control of ram velocity in injection molding machines. *Engineering Applications of Artificial Intelligence*, 15(1):65–72, 2002.

[446] K. K. Tan and S. Zhao. Iterative reference adjustment for high precision and repetitive motion control applications. In *Proceedings of the 2002 IEEE International Symposium on Intelligent Control*, pages 131–136, Vancouver, Canada, Oct. 2002.

[447] Andrew S. Tanenbaum. *Compuer Networks*. Prentice Hall, Upper Saddle River, NJ, 1996.

[448] A. Tayebi. Adaptive iterative learning control for robot manipulators. In *Proceedings of the 2003 American Control Conference*, pages 4518–4523, Denver, CO, June 2003.

[449] A. Tayebi. Adaptive iterative learning control for robot manipulators. *Automatica*, 40(7):1195–1203, 2004.

[450] A. Tayebi and J. X. Xu. Observer-based iterative learning control for a class of time-varying nonlinear systems. *IEEE Trans. on Circuits and Systems I – Fundamental Theory and Applications*, 50(3):452–455, 2003.

[451] A. Tayebi and M. B. Zaremba. Iterative learning control for non-linear systems described by a blended multiple model representation. *Int. J. of Control*, 75(16–17):1376–1384, 2002.

[452] A. Tayebi and M. B. Zaremba. Robust ILC design is straightforward for uncertain LTI systems satisfying the robust performance condition. In *Proceedings of IFAC 15th World Congress*, Barcelona, Spain, July 2002. IFAC.

[453] A. Tayebi and M.B. Zaremba. Internal model-based robust iterative learning control for uncertain LTI systems. In *Proceedings of the 39th IEEE Conference on Decision and Control*, pages 3439–3444, Sydney, Australia, Dec. 2000.

[454] S. P. Tian, S. L. Xie, and Y. L. Fu. A nonlinear algorithm of iteration learning control based on analysis of vector charts. *Dynamics of Continuous Discrete and Impulsive Systems – Series B – Applications & Algorithms*, pages 154–161, 2003.

[455] Yu-Ping Tian and Xinghuo Yu. Robust learning control for a class of uncertain nonlinear systems. In *Proceedings of IFAC 15th World Congress*, Barcelona, Spain, July 2002. IFAC.

[456] Yodyium Tipsuwan and Mo-Yuen Chow. Network-based controller adaptation based on QoS negotiation and deterioration. In *Proc. of the 27th Annual Conference of the IEEE Industrial Electronics Society*, 2001.

[457] R. Tousain, E. van der Meche, and O. Bosgra. Design strategy for iterative learning control based on optimal control. In *Proceedings of the 40th IEEE Conference on Decision and Control*, pages 4463–4468, Orlando, FL, Dec. 2001.

[458] T. Y. Townley and D. H. Owens. Output steering using iterative learning control. In *Proceedings of the 2001 European Control Conference*, Seminário de Vilar, Porto, Portugal, Sept. 2001. ECC.

[459] C. H. Tsai and C. J. Chen. Application of iterative path revision technique for laser cutting with controlled fracture. *Optics and Lasers in Engineering*, 41(1):189–204, 2004.

[460] Chin Tzong Pang, Yung Yih Lur, and Sy-Ming Guu. A new proof of Mayer's theorem. *Linear Algebra and Its Applications*, 350:273–278, 2002.

[461] M. Uchiyama. Formulation of high-speed motion pattern of a mechanical arm by trial. *Trans. SICE (Soc. Instrum. Contr. Eng.)*, 14(6):706–712(in Japanese), 1978.

[462] Job van Amerongen. Mechatronic design. In *Proceedings of the 3rd Mechatronics Forum International Conference*, Atlanta, Georgia, Sept. 2000.

[463] P. M. Lynch and R. Vangal. Tracking partially occluded two dimensional shapes. In *Proc. of the SPIE 1989*, 1989.

[464] W. J. R. Velthuis, T. J. A. de Vries, P. Schaak, and E. W. Gaal. Stability analysis of learning feed-forward control. *Automatica*, 36(12):1889–1895, 2000.

[465] M. H. A. Verwoerd, G. Meinsma, and T. J. A. de Vries. On the use of noncausal LTI operators in iterative learning control. In *Proceedings of the 41st IEEE Conference on Decision and Control*, pages 3362–3366, Las Vegas, Nevada, Dec. 2002.

[466] M. H. A. Verwoerd, G. Meinsma, and T. J. A. de Vries. On equivalence classes in iterative learning control. In *Proceedings of the 2003 American Control Conference*, pages 3632–3637, Denver, CO, June 2003.

[467] Mark Verwoerd. *Iterative learning control: a critical review*. PhD thesis, University of Twente (Netherlands), 2004.

[468] Victor Villagrán and Daniel Sbarbaro. A new approach for tuning MIMO PID controllers using iterative learning. In *Proceedings of the 14th World Congress of IFAC*, pages 247–252, Beijing, China, 1999. IFAC.

[469] V. J. Waissman, C. B. Youssef, and R. G. Vazquez. Iterative learning control for a fedbatch lactic acid reactor. In *Proceedings of the 2002 IEEE International Conference on Systems, Man and Cybernetics*, Oct. 2002.

[470] Gregory C. Walsh, Octavian Beldimna, and Linda Bushnell. Asymptotic behavior of networked control systems. In *Proc. of the 1999 IEEE International Conference on Control Applications*, Kohala Coast-Island, Hawaii, 1999.

[471] Gregory C. Walsh, Hong Ye, and Linda G. Bushnell. Stability analysis of networked control systems. *IEEE Trans. Control Systems Technology*, 10:438–446, 2002.

[472] D. W. Wang and C. C. Cheah. An iterative learning-control scheme for impedance control of robotic manipulators. *Int. J. of Robotics Research*, 17(10):1091–1104, 1998.

[473] Danwei Wang. On anticipatory iterative learning control designs for continuous time nonlinear dynamic systems. In *Proceedings of the 38th IEEE Conference on Decision and Control*, pages 1605–1610, Phoenix, AZ, Dec. 1999.

[474] Danwei Wang. On D-type and P-type ILC designs and anticipatory approach. *Int. J. of Control*, 73(10):890–901, 2000.

[475] D. W. Wang. Convergence and robustness of discrete time nonlinear systems with iterative learning control. *Automatica*, 34(11):1445–1448, 1998.

[476] Long Wang, Zhizhen Wang, and Wensheng Yu. Stability of polytopic polynomial matrices. In *Proceedings of American Control Conference*, pages 4695–4696, Arlington, Virginia, June 2001. ACC.

[477] Mingyan Wang, Ben Guo, Yudong Guan, and Hao Zhang. Design of electric dynamic load simulator based on recurrent neural networks.

In *Proceedings of the IEEE International Electric Machines and Drives Conference*, pages 207–210, June 2003.

[478] Qiang Wang, Qi-Rong Jiang, Ying-Duo Han, Jian-Xin Xu, and Xiao-Feng Bao. Advanced modeling and control package for power system problems. In *Proceedings of the IEEE 1999 International Conference on Power Electronics and Drive Systems*, pages 696–701, Hong Kong, July 1999.

[479] Y. C. Wang, C. J. Chien, and C. C. Teng. Direct adaptive iterative learning control of nonlinear systems using an output-recurrent fuzzy neural network. *IEEE Trans. on Systems Man and Cybernetics Part B – Cybernetics*, 34(3):1348–1359, 2004.

[480] Ying-Chung Wang, Chiang-Ju Chien, and Ching-Cheng Teng. A new recurrent fuzzy neural network based iterative learning control system. In *Proceedings of the 3rd Asian Control Conference*, Singapore, Sept. 2002. ASCC.

[481] Zhizhen Wang, Long Wang, and Wensheng Yu. Improved results on robust stability of multivariable interval control systems. In *Proceedings of American Control Conference*, pages 4463–4468, Denver, CO, June 4–6 2003. ACC.

[482] Zidong Wang, D. W. C. Ho, and Xiaohui Liu. Variance-constrained filtering for uncertain stochastic systems with missing measurements. *IEEE Trans. Automatic Control*, 48:1254–1258, 2003.

[483] T. Watabe, M. Yamakita, T. Mita, and M. Ohta. Output zeroing and iterative learning control for 3 link acrobat robot. In *Proceedings of the 41st SICE Annual Conference*, pages 2579–2584, Aug. 2002.

[484] Hao-Ping Wen. *Design of adaptive and basis function based learning and repetitive control*. PhD thesis, Columbia University, New York, NY, 2001.

[485] H. P. Wen, M. Q. Phan, and R. W. Longman. Bridging learning and repetitive control using basis functions. In *Advances in the Astronautical Sciences, Vol. 99, Part 1*, pages 335–354, 1998.

[486] Jan C. Willems. Least squares stationary optimal control and the algebraic Riccati equations. *IEEE Trans. on Automatic Control*, 16(6):621–634, 1971.

[487] Carl Wood, Shayne Rich, Monte Frandsen, Morgan Davidson, Russell Maxfield, Jared Keller, Braden Day, Matt Mecham, and Kevin Moore. Mechatronic design and integration for a novel omni-directional robotic vehicle. In *Proceedings of the 3rd Mechatronics Forum International Conference*, Atlanta, Georgia, Sept. 2000.

[488] Hansheng Wu, H. Kawabata, and K. Kawabata. Decentralized iterative learning control schemes for large scale systems with unknown interconnections. In *Proceedings of the 2003 IEEE Conference on Control Applications*, pages 1290–1295, CCA 2003, June 2003.

[489] Huaiyu Wu, Zhaoying Zhou, Shenshu Xiong, and Wendong Zhang. Adaptive iteration learning control and its applications for FNS multi-

joint motion. In *Proceedings of the 17th IEEE Instrumentation and Measurement Technology Conference*, pages 983–987, Baltimore, MD, May 2000. IMTC 2000.

[490] Jizhong Xiao, Qing Song, and Danwei Wang. A learning control scheme based on neural networks for repeatable robot trajectory tracking. In *Proceedings of the 1999 IEEE International Symposium on Intelligent Control/Intelligent Systems and Semiotics*, pages 102–107, Cambridge, MA, Sept. 1999.

[491] Lei Xie, Jian-Ming Zhang, and Shu-Qing Wang. Stability analysis of networked control system. In *Proc. of the First International Conference on Machine Learning and Cybernetics*, Beijing, 2002.

[492] Jan-Xin Xu and Ying Tan. On the robust optimal design and convergence speed analysis of iterative learning control approaches. In *Proceedings of IFAC 15th World Congress*, Barcelona, Spain, July 2002. IFAC.

[493] Z. Xiong and J. Zhang. Batch-to-batch optimal control of nonlinear batch processes based on incrementally updated models. *IEE Proceedings Part D – Control Theory and Applications*, 151(2):158–165, 2004.

[494] Z. H. Xiong and J. Zhang. Product quality trajectory tracking in batch processes using iterative learning control based on time-varying perturbation models. *Industrial & Engineering Chemistry Research*, 42(26):6802–6814, 2003.

[495] J. X. Xu, Y. Q. Chen, T. H. Lee, and S. Yamamoto. Terminal iterative learning control with an application to RTPCVD thickness control. *Automatica*, 35(9):1535–1542, 1999.

[496] J. X. Xu, Q. P. Hu, T. H. Lee, and S. Yamamoto. Iterative learning control with Smith time delay compensator for batch processes. *J. of Process Control*, 11(3):321–328, 2001.

[497] J. X. Xu, T. H. Lee, and Y. Tan. Enhancing trajectory tracking for a class of process control problems using iterative learning. *Engineering Applications of Artificial Intelligence*, 15(1):53–64, 2002.

[498] J. X. Xu, T. H. Lee, and H. W. Zhang. Analysis and comparison of iterative learning control schemes. *Engineering Applications of Artificial Intelligence*, 17(6):675–686, 2004.

[499] J. X. Xu, S. K. Panda, Y. J. Pan, T. H. Lee, and B. H. Lam. Improved PMSM pulsating torque minimization with iterative learning and sliding mode observer. In *Proceedings of the 26th Annual Conference of the IEEE Industrial Electronics Society*, pages 1931–1936, Nagoya Japan, Oct. 2000.

[500] J. X. Xu, S. K. Panda, Y. J. Pan, T. H. Lee, and B. H. Lam. A modular control scheme for PMSM speed control with pulsating torque minimization. *IEEE Trans. on Industrial Electronics*, 51(3):526–536, 2004.

[501] J. X. Xu and Y. Tan. A composite energy function-based learning control approach for nonlinear systems with time-varying parametric uncertainties. *IEEE Trans. on Automatic Control*, 47(11):1940–1945, 2002.

[502] J. X. Xu and Y. Tan. On the P-type and Newton-type ILC schemes for dynamic systems with non-affine-in-input factors. *Automatica*, 38(7):1237–1242, 2002.

[503] J. X. Xu and Y. Tan. Robust optimal design and convergence properties analysis of iterative learning control approaches. *Automatica*, 38(11):1867-1880, 2002.

[504] J. X. Xu and Y. Tan. Analysis and robust optimal design of iteration learning control. In *Proceedings of the 2003 American Control Conference*, pages 3038–3043, Denver, CO, June 2003.

[505] J. X. Xu, Y. Tan, and T. H. Lee. Iterative learning control design based on composite energy function with input saturation. *Automatica*, 40(8):1371–1377, 2004.

[506] J. X. Xu and J. Xu. On iterative learning from different tracking tasks in the presence of time-varying uncertainties. *IEEE Trans. on Systems, Man and Cybernetics Part B – Cybernetics*, 34(1):589–597, 2004.

[507] J. X. Xu and R. Yan. Fixed point theorem-based iterative learning control for LTV systems with input singularity. *IEEE Trans. on Automatic Control*, 48(3):487–492, 2003.

[508] J. X. Xu and R. Yan. Iterative learning control design without a priori knowledge of the control direction. *Automatica*, 40(10):1803–1809, 2004.

[509] Jian-Xin Xu and Ji Qian. New ILC algorithms with improved convergence for a class of non-affine functions. In *Proceedings of the 37th IEEE Conference on Decision and Control*, pages 660–665, Tampa, FL, Dec. 1998.

[510] Jian Xin Xu, Tong Heng Lee, and Hou Tan. Enhancing trajectory tracking for a class of process control problems using iterative learning. In *Proceedings of the 3rd Asian Control Conference*, Shanghai, China, 2000. ASCC.

[511] Jian-Xin Xu, Tong Heng Lee, Jing Xu, Qiuping Hu, and S. Yamamoto. Iterative learning control with Smith time delay compensator for batch processes. In *Proceedings of the 2001 American Control Conference*, pages 1972–1977, Arlington, VA, June 2001.

[512] Jian-Xin Xu, Tong Heng Lee, and Heng-Wei Zhang. Comparative studies on repeatable runout compensation using iterative learning control. In *Proceedings of the 2001 American Control Conference*, pages 2834–2839, Arlington, VA, June 2001.

[513] Jian-Xin Xu, Tong Heng Lee, and Heng-Wei Zhang. On the ILC design and analysis for a HDD servo system. In *Proceedings of IFAC 15th World Congress*, Barcelona, Spain, July 2002. IFAC.

[514] Jian-Xin Xu, Kok-Hwa Liew, and Chang-Chieh Hang. Comparative studies of P and PI type iterative learning control algorithms for a class of process control. In *Proceedings of the 14th World Congress of IFAC*, pages 165–170, Beijing, China, 1999. IFAC.

[515] Jian-Xin Xu and Ying Tan. New iterative learning control approaches for nonlinear non-affine MIMO dynamic systems. In *Proceedings of the*

2001 American Control Conference, pages 896–901, Arlington, VA, June 2001.

[516] Jian-Xin Xu and Ying Tan. On the convergence speed of a class of higher-order ILC schemes. In *Proceedings of the 40th IEEE Conference on Decision and Control*, pages 4932–4937, Orlando, FL, Dec. 2001.

[517] Jian-Xin Xu and Ying Tan. On the robust optimal design and convergence speed analysis of iterative learning control approaches. In *Proceedings of IFAC 15th World Congress*, Barcelona, Spain, July 2002. IFAC.

[518] Jian-Xin Xu and Ying Tan. *Linear and Nonlinear Iterative Learning Control*. Lecture Notes in Control and Information Sciences. Springer, New York, 2003.

[519] Jian-Xin Xu, Ying Tan, and Tong-Heng Lee. Iterative learning control design based on composite energy function with input saturation. In *Proceedings of the 2003 American Control Conference*, pages 5129–5134, Denver, CO, June 2003.

[520] Jian-Xin Xu and Badrinath Viswanathan. On the integrated learning control method. In *Proceedings of the 14th World Congress of IFAC*, pages 495–500, Beijing, China, 1999. IFAC.

[521] Jian-Xin Xu and Badrinath Viswanathan. Recursive direct learning of control efforts for trajectories with different magnitude scales. In *Proceedings of the 3rd Asian Control Conference*, Shanghai, China, 2000. ASCC.

[522] Jian-Xin Xu, Badrinath Viswanathan, and Zhihua Qu. Robust learning control for robotic manipulators with an extension to a class of non-linear systems. *Int. J. of Control*, 73(10):858–870, 2000.

[523] Jian-Xin Xu and Wai Yen Wong. Comparative studies on neuro-assisted iterative learning control schemes. In *Proceedings of the 14th World Congress of IFAC*, pages 453–458, Beijing, China, 1999. IFAC.

[524] Jian-Xin Xu and Jing Xu. Iterative learning control for non-uniform trajectory tracking problems. In *Proceedings of IFAC 15th World Congress*, Barcelona, Spain, July 2002. IFAC.

[525] Jian-Xin Xu, Jing Xu, and Badrinath Viswanathan. Recursive direct learning of control efforts for trajectories with different magnitude scales. *Asian Journal of Control*, 4(1):49–59, 2002.

[526] Jian-Xin Xu and Rui Yan. Fixed point theorem based iterative learning control for LTV systems with input singularity. In *Proceedings of the 2003 American Control Conference*, pages 3655–3660, Denver, CO, June 2003.

[527] Jian-Xin Xu and Rui Yan. Iterative learning control design without a priori knowledge of control directions. In *Proceedings of the 2003 American Control Conference*, pages 3661–3666, Denver, CO, June 2003.

[528] Jing Xu and Jian-Xin Xu. Memory based nonlinear internal model: What can a control system learn. In *Proceedings of the 3rd Asian Control Conference*, Singapore, 2002. ASCC.

[529] J.X. Xu and Z.H. Qu. Robust iterative learning control for a class of nonlinear systems. *Automatica*, 34(8):983–988, 1998.

[530] J.X. Xu and B. Viswanathan. Adaptive robust iterative learning control with dead zone scheme. *Automatica*, 36(1):91–99, 2000.

[531] W. Xu, C. Shao, and T. Chai. Learning control for a class of constrained mechanical systems with uncertain disturbances. In *Proceedings of the 1997 American Control Conference*, Albuquerque, NM, June 1997.

[532] Jian-Xin Xu and Wen-Jun Cao. Synthesized learning variable structure control approaches for repeatable tracking control tasks. In *Proceedings of the 14th World Congress of IFAC*, Beijing, China, 1999. IFAC.

[533] M. Yamada, Li Xu, and O. Saito. Iterative learning control of a robot manipulator using n-D practical tracking approach. In *Proceedings of the 47th Midwest Symposium on Circuits and Systems*, pages II–565–II–568, July 2004.

[534] M. Yamakita, M. Ueno, and T. Sadahiro. Trajectory tracking control by an adaptive iterative learning control with artificial neural networks. In *Proceedings of the 2001 American Control Conference*, pages 1253–1255, Arlington, VA, June 2001.

[535] M. Yamakita, T. Yonemura, Y. Michitsuji, and Z. Luo. Stabilization of acrobat robot in upright position on a horizontal bar. In *Proceedings of the IEEE International Conference on Robotics and Automation*, pages 3093–3098, May 2002.

[536] X. G. Yan, I. M. Chen, and J. Lam. D-type learning control for nonlinear time-varying systems with unknown initial states and inputs. *Trans. of the Institute of Measurement and Control*, 23(2):69–82, 2001.

[537] D. R. Yang, K. S. Lee, H. J. Ahn, and J. H. Lee. Experimental application of a quadratic optimal iterative learning control method for control of wafer temperature uniformity in rapid thermal processing. *IEEE Trans. on Semiconductor Manufacturing*, 16(1):36–44, 2003.

[538] Pai-Hsueh Yang. *Control design for a class of unstable nonlinear systems with input constraint*. PhD thesis, University of California, Berkeley, 1998.

[539] Shengyue Yang, Xiaoping Fan, and An Luo. Adaptive robust iterative learning control for uncertain robotic systems. In *Proceedings of the 4th World Congress on Intelligent Control and Automation*, pages 964–968, June 2002.

[540] Yi Yang. *Injection molding control: From process to quality*. PhD thesis, Hong Kong Univ. of Sci. and Tech. (China), 2004.

[541] Yongqiang Ye and Danwei Wang. Multi-channel design for ILC with robot experiments. In *Proceedings of the 7th International Conference on Control, Automation, Robotics and Vision*, pages 1066–1070, Dec. 2002.

[542] Yongqiang Ye and Danwei Wang. Better robot tracking accuracy with phase lead compensated ILC. In *Proceedings of the IEEE International*

Conference on Robotics and Automation, pages 4380–4385, Taipei, Taiwan, Sept. 2003.

[543] Rama K. Yedavalli. An improved extreme point solution for checking robust stability of interval matrices with much reduced vertex set and combinatorial effort. In *2002 American Control Conference*, pages 3902–3907, Arlington, VA, June 25-27 2001. AACC.

[544] Rama K. Yedavalli. It suffices to check only two special vertex matrices in Kronecker space to analyze the robust stability of an interval matrix. In *Proceedings of American Control Conference*, pages 1266–1271, Anchorage, AK, May 2002. ACC.

[545] H. Yu, M. Deng, T. C. Yang, and D. H. Owens. Model reference parametric adaptive iterative learning control. In *Proceedings of IFAC 15th World Congress*, Barcelona, Spain, July 2002. IFAC.

[546] Shao-Juan Yu, Suo-Lin Duan, and Ju-Hua Wu. Study of fuzzy learning control for electro-hydraulic servo control systems. In *Proceedings of the 2003 International Conference on Machine Learning and Cybernetics*, pages 591–595, Nov. 2003.

[547] Shao-Juan Yu, Ju-Hua Wu, and Xue-Wen Yan. A PD-type open-closed-loop iterative learning control and its convergence for discrete systems. In *Proceedings of the 2002 International Conference on Machine Learning and Cybernetics*, pages 659–662, Beijing, China, Nov. 2002.

[548] Lisong Yuan, Luke E. K. Achenie, and Weisun Jiang. Robust H_∞ control for linear discrete-time systems with norm-bounded time varying uncertainty. *Systems and Control Letters*, 27(4):199–208, 1996.

[549] Xiaoming Zha and Yunping Chen. The iterative learning control strategy for hybrid active filter to dampen harmonic resonance in industrial power system. In *Proceedings of the IEEE International Symposium on Industrial Electronics*, pages 848–853, June 2003.

[550] Xiaoming Zha, Jianjun Sun, and Yunping Chen. Application of iterative learning control to active power filter and its robustness design. In *Proceedings of the IEEE 34th Annual Conference on Power Electronics Specialist*, pages 785–790. PESC '03, June 2003.

[551] Xiaoming Zha, Qian Tao, Jianjun Sun, and Yunping Chen. Development of iterative learning control strategy for active power filter. In *Proceedings of the Canadian Conference on Electrical and Computer Engineering*, pages 240–245, May 2002.

[552] W. Zhang, M. S. Branicky, and S. M. Phillips. Stability of networked control systems. *IEEE Control Systems Magazine*, 21:84–99, 2001.

[553] Y. M. Zhang and R. Kovacevic. Robust control of interval plants: a time domain method. *IEE Proceedings – Control Theory and Applications*, 144(4):347–353, 1997.

[554] Danian Zheng. *Iterative learning control of an electrohydraulic injection molding machine with smoothed fill-to-pack transition and adaptive filtering*. PhD thesis, University of Illinois at Urbana-Champaign, 2002.

228   References

[555] Chi Zhu, Y. Aiyama, and T. Arai. Releasing manipulation with learning control. In *Proceedings of the 1999 IEEE International Conference on Robotics and Automation*, pages 2793–2798, Detroit, MI, May 1999.

Index